Adeno-Associated Viral Vectors for Gene Therapy

LABORATORY TECHNIQUES IN BIOCHEMISTRY AND MOLECULAR BIOLOGY

Series Editors

P.C. van der Vliet—*Department for Physiological Chemistry, University of Utrecht, Utrecht, The Netherlands*

and

S. Pillai—*MGH Cancer Center, Boston, Massachusetts, USA*

Volume 31

ELSEVIER

AMSTERDAM • BOSTON • HEIDELBERG • LONDON • NEW YORK • OXFORD
PARIS • SAN DIEGO • SAN FRANCISCO • SINGAPORE • SYDNEY • TOKYO

ADENO-ASSOCIATED VIRAL VECTORS FOR GENE THERAPY

Edited by

T.R. Flotte
Powell Gene Therapy Center
University of Florida
Gainesville, Florida, USA

K.I. Berns
College of Medicine
University of Florida
Gainesville, Florida, USA

2005

ELSEVIER

AMSTERDAM • BOSTON • HEIDELBERG • LONDON • NEW YORK • OXFORD
PARIS • SAN DIEGO • SAN FRANCISCO • SINGAPORE • SYDNEY • TOKYO

ELSEVIER B.V.	ELSEVIER Inc.	ELSEVIER Ltd.	ELSEVIER Ltd.
Sara Burgerhartstraat 25	525 B Street	The Boulevard	84 Theobalds Road
P.O. Box 211, 1000 AE	Suite 1900, San Diego	Langford Lane, Kidlington,	London WC1X 8RR
Amsterdam, The Netherlands	CA 92101-4495, USA	Oxford OX5 1GB, UK	UK

© 2005 Elsevier B.V. All rights reserved.

This work is protected under copyright by Elsevier B.V., and the following terms and conditions apply to its use:

Photocopying
Single photocopies of single chapters may be made for personal use as allowed by national copyright laws. Permission of the Publisher and payment of a fee is required for all other photocopying, including multiple or systematic copying, copying for advertising or promotional purposes, resale, and all forms of document delivery. Special rates are available for educational institutions that wish to make photocopies for non-profit educational classroom use.

Permissions may be sought directly from Elsevier's Rights Department in Oxford, UK; phone (+44) 1865 843830, fax (+44) 1865 853333, e-mail: permissions@elsevier.com. Requests may also be completed on-line via the Elsevier homepage (http://www.elsevier.com/locate/permissions).

In the USA, users may clear permissions and make payments through the Copyright Clearance Center, Inc., 222 Rosewood Drive, Danvers, MA 01923, USA; phone: (+1) (978) 7508400, fax: (+1) (978) 7504744, and in the UK through the Copyright Licensing Agency Rapid Clearance Service (CLARCS), 90 Tottenham Court Road, London W1P 0LP, UK; phone: (+44) 20 7631 5555; fax: (+44) 20 7631 5500. Other countries may have a local reprographic rights agency for payments.

Derivative Works
Tables of contents may be reproduced for internal circulation, but permission of the Publisher is required for external resale or distribution of such material. Permission of the Publisher is required for all other derivative works, including compilations and translations.

Electronic Storage or Usage
Permission of the Publisher is required to store or use electronically any material contained in this work, including any chapter or part of a chapter.

Except as outlined above, no part of this work may be reproduced, stored in a retrieval system or transmitted in any form or by any means, electronic, mechanical, photocopying, recording or otherwise, without prior written permission of the Publisher.
Address permissions requests to: Elsevier's Rights Department, at the fax and e-mail addresses noted above.

Notice
No responsibility is assumed by the Publisher for any injury and/or damage to persons or property as a matter of products liability, negligence or otherwise, or from any use or operation of any methods, products, instructions or ideas contained in the material herein. Because of rapid advances in the medical sciences, in particular, independent verification of diagnoses and drug dosages should be made.

First edition 2005

Library of Congress Cataloging in Publication Data
A catalog record is available from the Library of Congress.

British Library Cataloguing in Publication Data
A catalogue record is available from the British Library.

ISBN-13: 978-0-444-51949-8
ISBN-10: 0-444-51949-1
ISBN: 0-7204-4200-1 (Series)
ISSN: 0075-7535 (Series)

∞ The paper used in this publication meets the requirements of ANSI/NISO Z39.48-1992 (Permanence of Paper).
Printed in The Netherlands.

Preface

Adeno-associated viruses (AAV) are ubiquitous in the human population (>90% of adults are seropositive) but have never been implicated as the cause of any disease. They were discovered in the 1960s as particles which contaminated preparations of adenovirus which had been considered to be "chemically pure" until observed in the electron microscope. Early studies of AAV were confined only to a few laboratories, because of the lack of apparent pathogenicity-related interest in the virus. AAV was considered to be defective because productive infection required coinfection with a helper adenovirus. However, very early on M.D. Hoggan discovered that AAV caused persistent infection and because AAV was the only nuclear DNA virus which could readily be maintained in the latent phase in cell culture (by simply omitting the helper adenovirus), it became a good model for the study of persistent infection at the molecular level. Thus, AAV became the archetypal viral parasite by virtue of its ability to have the cell take over the task of perpetuating the viral genome, while not harming the host at either the level of the cell or the intact host. Even more interestingly, AAV has been suggested to have a protective role for the host. Several epidemiological studies have found that women who suffer from cervical carcinoma are remarkably negative for serological evidence of prior AAV infection. We now know that several types of human papillomaviruses are the likely cause of cervical carcinoma, that AAV is inhibitory for papillomavirus replication, and that the female genital tract is a common site of AAV persistent infection. Thus, AAV is the first human virus which has been implicated as a possible symbiont which has a positive effect on the host.

There are two general requirements for human gene therapy. The first is a cloned, corrective gene; many of these are now available. The second requirement is a safe, effective vehicle for introducing the corrective gene into the target tissues. The latter has been the major challenge to successful gene transfer. Because viruses naturally introduce genes into cells and cause them to be expressed, they have been intensively studied as possible vectors. For most uses foreseen, an ideal vector would introduce a transgene which would then be able to be expressed for an extended period of time (ideally for a lifetime in many instances), without any toxic side effects of the therapy. Since most viruses are pathogens, reduction of pathogenicity has been of major concern. AAV apparently is not a natural pathogen. Current AAV vectors contain no viral genes and so do not elicit a significant immune response after a single administration and, possibly for this reason, allow extended transgene expression. As a consequence, AAV has gained increasing favor as a potential vector for human gene therapy and this volume is a summary of the current state of AAV vectors.

Terrence R. Flotte

Contents

Preface v

Chapter 1.   Adeno-associated viral vectors for gene therapy 1
T. R. Flotte and K. I. Berns

    1.1.   Biological properties of adeno-associated virus 1
    1.2.   AAV-based gene therapy vectors 2
    1.3.   In vivo applications of rAAV 4
    1.4.   Clinical experience with rAAV 7
    1.5.   Persistence of rAAV vectors 8
    1.6.   Safety of rAAV vector delivery 11
    1.7.   Host range, alternate serotypes, and capsid modifications 12
    1.8.   Remaining questions 13
           References 13

Chapter 2.   Production of research and clinical-grade recombinant adeno-associated virus vectors 19
J. D. Francis and R. O. Snyder

    2.1.   Adeno-associated virus biology 20
    2.2.   rAAV preparation 24
    2.3.   Protocols 27
    2.4.   Small-scale rAAV 1, 2, and 5 vector purification 28
    2.5.   Large-scale rAAV purification 30
    2.6.   rAAV vector characterization 31
    2.7.   Safety testing 34
    2.8.   Pre-clinical regulatory compliance activities 35
    2.9.   Manufacture of clinical-grade rAAV vectors 38
    2.10.   Clinical manufacturing regulatory compliance activities 44
           References 49

Chapter 3.   Gene therapy for hemophilia................. 57
C. Mah

    3.1.   Non-viral DNA vectors............................. 59
    3.2.   Adenovirus vectors............................... 60
    3.3.   Retrovirus vectors................................ 62
    3.4.   Adeno-associated virus vectors 64
    3.5.   Immunological considerations 69
    3.6.   Laboratory protocols 72
           References...................................... 73

*Chapter 4.   Recombinant AAV vectors for gene transfer
to the lung: a compartmental approach*................... 83
T. R. Flotte

    4.1.   Introduction..................................... 83
    4.2.   Genes, targets and vectors for the lung................ 84
    4.3.   Therapies targeting the alveoli...................... 85
    4.4.   Therapies targeting the airways 87
    4.5.   Therapies targeting the pulmonary vasculature and pleura..... 92
    4.6.   Future directions 93
           References...................................... 94

*Chapter 5.   Adeno-associated virus mediated gene therapy
for vascular retinopathies*............................. 103
B. J. Raisler, W.-T. Deng, K. I. Berns and W. W. Hauswirth

    5.1.   Introduction.................................... 103
    5.2.   New strategies for treating NV 106
    5.3.   Protocols...................................... 114
    5.4.   Discussion 118
           References..................................... 119

Chapter 6. Gene therapy for prevention and treatment of type 1 diabetes... 125
M. H. Kapturczak, B. R. Burkhardt and M. A. Atkinson

6.1.	The clinical problem diabetes	125
6.2.	Transplantation	126
6.3.	Allograft rejection: mechanisms for increasing graft acceptance.................................	127
6.4.	Recurrent autoimmunity as a mechanism of β cell allograft failure	129
6.5.	Gene transfer into islet cells	131
6.6.	Potential utility of rAAV-mediated gene therapy for islet transplantation and prevention of autoimmunity recurrence in type 1 diabetes.........................	136
6.7.	Progress in insulin replacement strategies utilizing gene therapy	142
6.8.	Summary and future directions	146
	References.......................................	147

Chapter 7. Gene therapy for kidney diseases............. 161
S. Chen, K. M. Madsen, C. C. Tisher and A. Agarwal

7.1.	Structure–function correlations	162
7.2.	Vector systems for gene delivery	163
7.3.	Methods of gene delivery	170
7.4.	Targeting specific cells in the kidney	172
7.5.	Application of gene therapy for specific kidney diseases...................................	176
	References.......................................	185

Chapter 8. AAV for disorders of the CNS 193
C. Burger, R. J. Mandel and N. Muzyczka

8.1.	Introduction	193
8.2.	Parkinson disease (PD).............................	198
8.3.	Alzheimer's disease (AD)	204
8.4.	Epilepsy...	206

8.5.	Lysosomal storage disorders (LSD)	208
8.6.	Conclusion	212
	References	213

Chapter 9. Gene therapy for cardiovascular applications .. 225
C. A. Pacak, C. Mah and B. J. Byrne

9.1.	Viral gene delivery systems	227
9.2.	Non-viral gene delivery systems	232
9.3.	Gene delivery route	233
9.4.	Cellular and gene therapy combinations	234
9.5.	Conclusions	234
9.6.	Methods	235
	References	239

Chapter 10. Gene therapy for lysosomal storage disorders. 243
K. O. Cresawn and B. J. Byrne

10.1.	The lysosome	243
10.2.	Lysosomal storage diseases	244
10.3.	Current therapies	244
10.4.	Gene therapy	245
10.5.	Glycogen storage disease type II	248
10.6.	Gene therapy for GSD II: Proof of concept studies	254
10.7.	Recombinant adeno-associated virus vector studies	256
10.8.	Recombinant AAV-mediated treatment of GSDII	257
10.9.	Gene therapy for CNS pathologies in LSDs	262
10.10.	Conclusion	265
	References	266

Index 277

CHAPTER 1

Adeno-associated viral vectors for gene therapy

Terence R. Flotte and Kenneth I. Berns

Department of Pediatrics, Powell Gene Therapy Center, Genetics Institute, University of Florida College of Medicine, Box 100296, Gainesville, FL 32610-0296, USA

1.1. Biological properties of adeno-associated virus

Adeno-associated viruses (AAV) are small DNA viruses that are known to infect many vertebrate species (Berns, 1990; Berns and Linden, 1995). The viral particle is icosahedral with a diameter of ~25 nm, non-enveloped, and contains a linear, single-stranded DNA genome of 4.6–4.8 kb. The virions are unusual because half contain the positive DNA strand and the rest contain a negative polarity strand. Although close to 90% of adult humans are seropositive for one or more of the over 100 primate types, no human diseases have been shown to be associated with viral infection. Infection of healthy cells in culture does not lead to viral replication, rather the DNA genome is uncoated, converted to the duplex form, and integrated into a specific site in the human genome (e.g., 19q13.3-qter for AAV2) so that a cryptic, latent infection ensues (Kotin et al., 1990, 1991, 1992). Productive AAV infection requires either a coinfection or superinfection with a helper virus, either an adeno- or a herpesvirus or exposure of the infected cells to a genotoxic agent such as ultraviolet irradiation.

The AAV genome is a linear DNA that contains two open reading frames (orf) (Hermonat et al., 1984; Tratschin et al., 1984). The orf in the right half of the genome encodes the three overlapping coat proteins. The three proteins result from alternative splicing and use of an alternative ACG initiator codon for translation of VP2 (Trempe and Carter, 1988). The orf in the left half of the genome encodes four regulatory Rep proteins, also with partially overlapping sequences (Mendelson et al., 1986). There are two transcripts starting at promoters at map positions 5 and 19 and both spliced and unspliced forms of both transcripts are translated to yield the four Rep proteins (Rep 78, 68, 52, and 40, respectively). The two larger Rep proteins (78, 68) regulate all phases of the AAV life cycle. Under intracellular conditions leading to latency, Rep inhibits viral gene expression and DNA replication, and is required for site-specific integration (Beaton et al., 1989; Kearns et al., 1996). When the conditions are permissive for productive infection, Rep is required for viral gene expression (Shi et al., 1991) and DNA replication, as well as rescue from the integrated state. Rep can bind to double-stranded DNA in a sequence-specific manner is both a DNA and DNA:RNA helicase and a sequence-specific DNA nickase (Im and Muzyczka, 1990). The genome has an inverted terminal repeat (itr) of 125 bases. The itr can fold on itself to form a T-shaped structure in which all but seven of the bases are paired: The seven unpaired bases include three at each of the fold back regions of the cross arms of the T and one unpaired base between the two cross arms. Each of the unpaired bases is an A or a T. The itr binds Rep, is an enhancer of gene expression, and is required for DNA replication and its negative regulation under non-permissive conditions (Pereira et al., 1997). Finally, the itr is required for encapsidation of the viral genome and site-specific integration.

1.2. AAV-based gene therapy vectors

AAV has been considered as a potential vector for gene therapy because of its wide range of tissue specificity, its ability to infect and be expressed in non-dividing cells and because of its lack of pathogenicity (Hermonat and Muzyczka, 1984; Tratschin et al., 1984). Generally the transgene is inserted between two itrs with deletion of both viral orfs. In this form a transgene of up to 4.7 kb can be inserted. There are several consequences to deletion of the *rep* gene. These include the loss of site-specificity of integration, although the ability to integrate is retained. Rep is a potent effector of gene expression and, thus, deletion removes the possibility of any toxicity due to this cause.

Production of vector virions requires transfection of the vector construct in plasmid form together with the introduction on separate plasmids and/or helper viruses of both the AAV genes and the helper virus genes required for a productive AAV infection (Xiao et al., 1998) (Fig. 1.1). Variations have included producing vectors

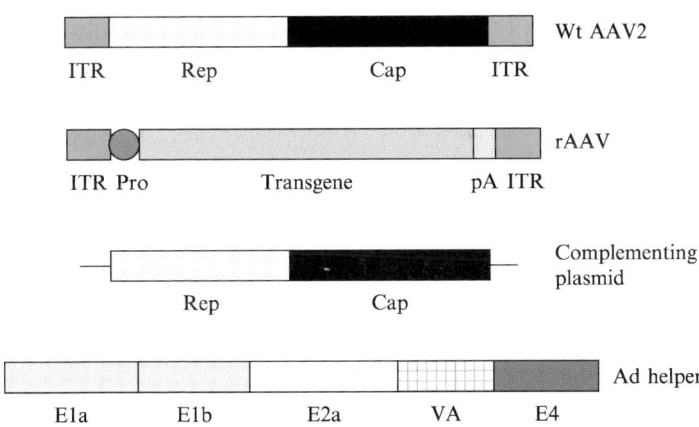

Fig. 1.1. The rAAV2 vector packaging scheme. The genome of AAV serotype 2 is shown in the top diagram. "itr" indicates the inverted terminal repeat sequences. Rep and Cap genes are shown. In the second panel from the top, a typical rAAV2 vector genome is shown with a heterologous promoter (Pro) and polyadenylation signal (pA) flanking the transgene, such as CFTR. In the third diagram is depicted a complementing plasmid that expresses Rep and Cap proteins within the packaging cell to activate vector genome replication and encapsidation. On the bottom are shown adenoviral gene products that also must be supplied *in trans* for efficient replication and packaging.

in host cells with one or both of the AAV genes contained in the cellular genome. The goal of vector production is high titer vector virions with no contamination by either wild-type AAV or intact helper virus. Current production protocols do permit these results and new techniques of fractionation preserve the viability of the vectors (Zolotukhin et al., 1999).

1.3. In vivo applications of rAAV

The development of more efficient systems for packaging and purification of wild-type-free preparations of rAAV has made it possible to test the safety, efficiency, and duration of transduction in a wide range of mammalian organs and tissues. While early in vitro studies questioned whether rAAV would be capable of gene transfer in nondividing cells, in vivo studies have indicated that rAAV mediates very efficient transduction of a wide range of terminally differentiated cells, including neurons, retinal photoreceptor cells, myofibers, bronchial epithelial cells, and hepatocytes (Flotte et al., 1993; Kaplitt et al., 1994; Kessler et al., 1996; Xiao et al., 1996; Clark et al., 1997; Flannery et al., 1997; Snyder et al., 1997; Klein et al., 1998).

The routes of delivery of rAAV in animal models have been largely based upon the specific needs dictated by the disease process to be treated. For example, rAAV-cystic fibrosis transmembrane conductance regulator (CFTR) vectors that have been developed for treatment of cystic fibrosis (CF) were tested in New Zealand white rabbits and in rhesus macaques by the endobronchial route (Flotte et al., 1993; Afione et al., 1996; Conrad et al., 1996). In each instance aliquots of vector were instilled directly into the lumen of a bronchus through a fiberoptic bronchoscope. Vector DNA transfer and mRNA expression were detectable (albeit at low levels) for more than 6 months in each instance, without any indication of inflammation or any other toxicity. Studies in rhesus monkeys also indicated that the likelihood of rescue of rAAV by concomitant wild-type AAV and adenovirus infection was low. These studies provided the pre-clinical data necessary for initiation of phase I clinical trials of rAAV in CF patients (Flotte et al., 1996; Wagner et al., 1998, 1999). More recently, numerous studies have indicated that rAAV transduction of skeletal muscle results in efficient expression of marker genes, secreted proteins, or gene products needed for correction of intrinsic muscle disease, such as delta sarcoglycan, one of the genes implicated in limb-girdle muscular dystrophy (Xiao et al., 2000). The efficiency of myofiber transduction has been remarkably high in both cardiac and skeletal muscle, and the duration of expression in rodent and canine models has been lifelong. Transduction of skeletal muscle in non-human primates has also been demonstrated to be safe, efficient, and long-lasting (Zhou et al., 1998). The transduction of muscle has provided a surprisingly efficient platform for the expression and secretion of secreted proteins, such as erythropoietin (Kessler et al., 1996), leptin (Murphy et al., 1997), alpha 1-antitrypsin (Song et al., 2001a), and clotting factor IX (Herzog et al., 1997). The latter has also been examined in early phase clinical trials in patients with hemophilia B (Kay et al., 2000).

The transduction of neurons of the brain and spinal cord by rAAV has also been remarkably efficient in a number of rodent and non-human primate models (Kaplitt et al., 1994; Peel et al., 1997; During et al., 1998; Klein et al., 1998). Stereotactic injection of rAAV into the substantia nigra or striatum has been used to deliver potentially therapeutic genes in animal models of Parkinson Disease (Kaplitt et al., 1994; During et al., 1998; Klein et al., 1998). These studies have shown efficient transgene expression with functional correction in both rodent and primate models. While the transduction of localized areas of the brain or spinal cord appears to be readily practical, the delivery of rAAV to wider regions of the brain, that may be needed for patients with some metabolic disorders, remains challenging.

Retinal photoreceptors are essentially neuron-like cells derived from neurectoderm, and they are like neurons in being very efficiently transduced by rAAV (Zolotukhin et al., 1996; Flannery et al., 1997). The delivery of rAAV by subretinal injection provides a direct means for delivering vector into a confined space where it can directly contact the photoreceptor layer and spread laterally across the retina. This anatomical arrangement and the excellent permissiveness of these cells for rAAV transduction has led to some very impressive results in several disease models, including a canine model of Leber's congenital amaurosis (Acland et al., 2001). In the latter study, a single subretinal injection of a rAAV vector into dogs with this genetic defect in the *rpe65* locus resulted in long-term restoration of electrophysiologic activity, pupillary response, and clinically evident visual acuity. The potential uses of rAAV-retinal transduction extend far beyond inborn errors, however. Photoreceptors, like skeletal myofibers, could be used to express proteins for local secretion, such as angiostatin and PEDF for the suppression of new vessel growth in models of diabetic retinopathy.

The transduction of hepatocytes with rAAV has an even broader range of potential applications. In general, hepatocytes are quite efficient for the secretion of therapeutic proteins, and have shown some advantage over skeletal muscle for the production of factor IX or alpha 1-antitrypsin in mouse models (Snyder et al., 1999; Song et al., 2001b) (Fig. 1.2). The transduction of hepatocytes with rAAV has also been sufficiently robust to achieve correction of a mouse model of phenylketonuria (Laipis et al., 2001). In that case a single portal vein injection of over 5×10^{12} infectious units of rAAV was able to cause a long-term decrease in the serum phenylalanine levels and a reversal of the coat color defect that results from impaired melanin production in this disease. The correction of intrinsic hepatocyte defects is likely to be even more challenging, however, since long-term transduction by rAAV2 appears to be limited to approximately 5% of hepatocytes (Miao et al., 1998, 2000). The safety of rAAV transduction in the liver has been quite good. In one anecdotal report hepatocellular carcinomas in animals treated with rAAV as neonates were shown NOT to be related to rAAV integration, since most of the tumors did not have any detectable rAAV vector sequences and those with detectable

Ch. 1 ADENO-ASSOCIATED VIRAL VECTORS FOR GENE THERAPY

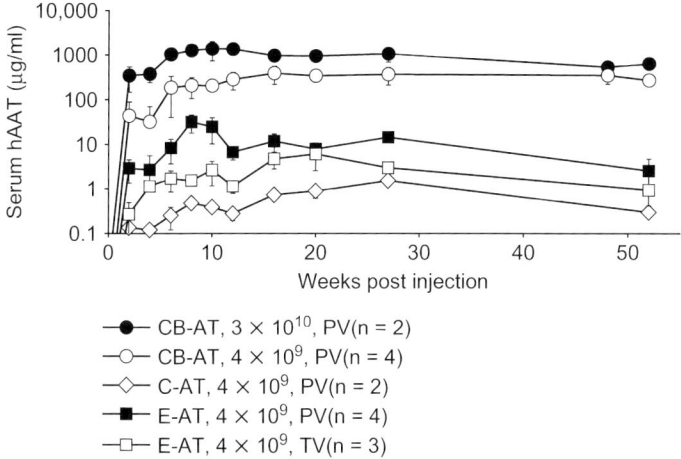

Fig. 1.2. Long-term expression of a rAAV2 vector after delivery to mouse liver. Aliquots of the indicated dosages of a CMV enhancer/chicken beta actin hybrid promoter-driven human alpha 1-antitrypsin (AAT)-expressing rAAV2 vector (CB-AT) or of a similar elongation factor 1-alpha promoter-driven vector (E-AT) were injected into C57Bl/6 mice either by portal vein (PV) or tail vein (TV). The serum levels of human AAT were measured by ELISA serially from 0 to 52 weeks after injection, and these values are shown on the y-axis.

sequences did not have evidence of genomic integration of the vector (Donsante et al., 2001).

1.4. Clinical experience with rAAV

To date, only three rAAV vectors have been used in human, although two of the three have been used in a number of trials utilizing several different routes of administration. The results have generally been quite consistent with those seen in animal models, in terms of the safety, efficiency, and duration of transgene expression. The rAAV-CFTR vector mentioned above has been used in four trials, including a trial of sequential nasal and bronchoscopic instillation into a single lung lobe, two trials of installation in the maxillary sinuses, and one additional trial of aerosol delivery (Flotte et al., 1996; Wagner et al., 1998, 1999; Virella-Lowell et al., 2000). Altogether vector administration has been safe in the over 70 CF patients who have been treated. The efficiency of transduction has varied somewhat. Transduction of the paranasal sinuses has been quite efficient, up to a maximal DNA efficiency of one copy per cell. This level of transduction was also associated with a partial correction of the epithelial electrolyte transport and a trend toward a decrease in pro-inflammatory cytokines, which are typically elevated in CF airway epithelia. Limitations to airway epithelial transduction have also been studied in these trials. These limitations include the inactivation of rAAV by products of CF-related lung inflammation and the relative paucity of high-affinity receptors on the apical surface of bronchial epithelial cells.

Two phase I trials have been undertaken with rAAV-factor IX (FIX) vectors, one involving intramuscular administration (Kay et al., 2000), and the second involving direct delivery to the liver by way of a catheter in the hepatic artery. Early results from the muscle delivery trial were quite encouraging, but the switch to the intrahepatic route was made in an attempt to exploit the more efficient secretory ability of hepatocytes. One additional trial was initiated using a rAAV-alpha sarcoglycan vector in one patient with limb-girdle muscular dystrophy. This last trial was halted after one patient due to factors unrelated to the study results.

1.5. Persistence of rAAV vectors

In most in vitro and in vivo applications of rAAV, it has been observed that rAAV vector genomes persist long-term, in most cases for the life span of the infected cell. This persistence may be viewed as a modified form of latent AAV infection. Given the pivotal role played by the Rep proteins in the site-specific integration of wild-type AAV, it is not surprising that Rep-deleted rAAV genomes behave somewhat differently than their wild-type counterparts (Kearns et al., 1996). The conversion of single-stranded rAAV genomes into duplex forms is slower than with wt-AAV, requiring anywhere from 10 days to 8 weeks in the various cell types (Song et al., 1998; Afione et al., 1999). There is evidence to support leading strand synthesis as the mode of conversion to duplex DNA in muscle fibers and bronchial epithelial cells, and evidence to support annealing of opposite strands of single-stranded vector DNA in hepatocytes (Fisher et al., 1996; Nakai et al., 2000). An understanding of the kinetics of the generation of duplex forms is crucial for assessment of the efficiency of transgene expression with rAAV vectors, since the presence of such forms is required for transcription. Early termination of an experiment utilizing a rAAV vector could lead to a significant underestimation of the potential efficiency of gene delivery.

Another aspect of rAAV latency that is unique is the relatively high abundance of persistent episomal forms and the lack of site-specific integration. The vast majority of rAAV genomes that persist in bronchial epithelium, muscle, and liver are episomal (Afione et al., 1996; Song et al., 1998, 2001a, b). The structure of these episomal forms varies and can include circular, linear, and large concatemeric species. The higher molecular weight forms may be generated by concatemerization of two or more vector genomes into a single larger molecule. This phenomenon, at least in murine skeletal muscle, appears to be dependent upon the function of the DNA-dependent protein kinase (DNA-PK) a key double-stranded DNA ligase responsible for non-homologous end-joining in the host cell (Song et al., 2001a).

The important contribution of these episomal forms to transgene expression has recently been demonstrated in mouse liver (Nakai et al., 2001). In this study a partial hepatectomy was performed on mice after portal vein injection of a rAAV vector, in order to stimulate proliferation of the transduced liver. In this situation one would predict that expression from *integrated* vector genomes would not decrease significantly, since the genome copy number would re-expand along with the repopulating hepatocyte population. In fact, both the vector DNA copy number and the expression level decreased to 10% of the original level, indicating that episomal vector DNA forms had been diluted out as the hepatocyte population expanded.

The ability of rAAV genomes to concatemerize with each other has recently been exploited as a strategy to overcome the packaging limit of the rAAV virion, which represents one of the only inherent limitations of this system (Duan et al., 2000; Sun et al., 2000; Yan et al., 2000). Two different strategies have been employed. The first is to incorporate a large transgene into one rAAV vector with only minimal promoter elements, such as the itr itself or the minimal thymidine kinase promoter. This vector is then coinfected along with a second vector carrying a combination of very active enhancer elements. The mixed vector infection forms concatemers that can express up to 600 times more transgene product than the minimal promoter vector alone, and this expression persists in murine skeletal muscle over the long-term. The second strategy is to create a split intron pair of vector constructs. In this approach, the promoter, the 5' end of the gene coding sequence, the splice donor site, and the first half of the intron are in one vector, while the second half of the intron, the splice acceptor, the 3' end of the gene, and the polyadenylation signal are in the second vector. The latter strategy has a disadvantage in that only one of four possible combinations of vector genome orientations will be functional, while in the former scheme any of the four would work (Fig. 1.3).

The predominance of episomal persistence in latent rAAV infection has a number of important implications. First, it helps to explain why rAAV gene transfer has been particularly effective in terminally differentiated nonproliferating cell populations. In these cells episomal forms are quite stable and are not diluted out by host cell division. A second important aspect of episomal persistence is that this mode of persistence decreases the risk of insertional mutagenesis that might otherwise be associated with random integration of vector DNA into the host cell genome.

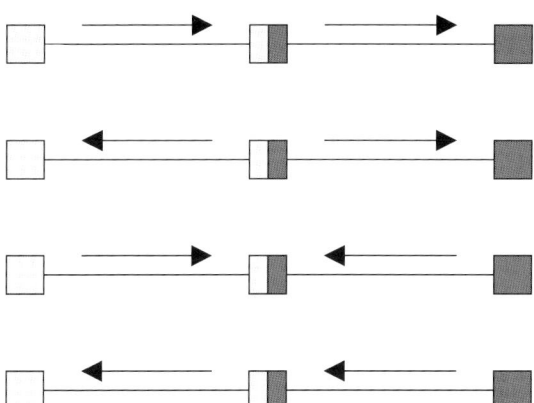

Fig. 1.3. Four possible orientations of products of intermolecular recombination. When one vector carries the entire transgene and the other an enhancer, all four will be active. When the two vectors carry the two halves of a single gene coding region with an intervening intron, only the first of these will be active.

1.6. Safety of rAAV vector delivery

One of the most remarkable aspects of rAAV-based gene delivery is the lack of any known toxicity from vector delivery. Other viral vectors elicit a wide range of toxicities immediately upon interaction with host cell receptors, including pro-inflammatory cytokine gene induction and promotion of humoral and cell-mediated immune responses. In contrast, rAAV vectors are remarkably benign. A wide range of pre-clinical studies and a limited number of clinical trials all indicate that there is no detectable toxicity from administration of rAAV, even at the highest deliverable doses used to date. Expression profiling of infected cells with microarray analysis indicates that host cell gene expression is remarkably unperturbed after rAAV infection. While some humoral immune responses are observed to the rAAV capsid, a series of studies indicates that cell-mediated immune responses to AAV are uncommon in the absence of adenovirus coinfection.

As mentioned above, rAAV has not been found to have significant genotoxicity, either. While there has been some theoretical concern that random-site integration from Rep-deleted rAAV transduction could result in insertional mutagenesis and possible carcinogenesis, this has not yet been observed. One anecdotal report of liver tumors in beta-glucuronidase (gusB)-deficient mice treated as newborns with intravenous injection of a rAAV vector expressing the gusB gene has recently been clarified (Donsante et al., in press). Careful analysis of the tumors demonstrated that vector DNA was not present in most tumors, indicating that they were more likely related to the underlying gusB deficiency or to the transgene, rather than to the vector itself. Furthermore, a detailed study of a large number of mice treated with other rAAV vectors with a very similar expression cassette has failed to show any evidence of inherent tumorigenicity from the vector.

1.7. Host range, alternate serotypes, and capsid modifications

The vast majority of studies with rAAV have been performed using AAV serotype 2-based vector genomes and capsids. The binding and entry of rAAV2 vectors depends upon specific interaction with an attachment receptor, which has been shown to be the heparan sulfate proteoglycan (HSP), and a co-receptor, which may be either the fibroblast growth factor receptor or the $alpha_v$-$beta_5$ integrin. It has been shown that certain cell types, such as ciliated bronchial epithelial cells and oligodendroglia are relatively less permissive for rAAV2 infection because of a paucity of these receptors.

Two strategies have been pursued to overcome this limitation and effectively broaden the cellular host range of rAAV vectors. The first is to utilize other rAAV serotypes, which may target other cellular receptors. Recent studies have shown an increased efficiency of gene delivery in lung, pancreatic islets, and glial cells with rAAV serotype 5 and an increase in the efficiency of delivery to muscle using rAAV serotype 1. The other approach has been to specifically target receptors either by use of bifunctional antibodies or by genetic incorporation of new peptide ligands into the AAV capsid sequence. Both strategies have been used successfully to enhance transduction of bronchial epithelial cells.

1.8. Remaining questions

AAV's unique life cycle and biological properties of AAV have been exploited with remarkable success in developing rAAV as a safe, effective, and persistent gene transfer vehicle in mammalian systems. While it seems likely that there will be clinical success with the existing rAAV technology, the most unique property of wt-AAV, its ability to undergo site-specific integration, has not yet been fully utilized. Transient Rep expression has been used to target rAAV integration to the AAVS1 site in a number of cell culture systems. If this process could be adapted for in vivo use, it could allow integration into a region that might allow for continued transgene gene expression whether or not the cell population was proliferating, while still minimizing the long-term risks of insertional mutagenesis. However, this approach presents its own challenges, since excessive or persistent Rep expression could either be cytotoxic or cytostatic, or might suppress transgene expression from the gene of interest. This suggests that tightly controlled Rep expression, such as that produced by wt-AAV itself through autoregulation of the p5 and p19 promoters, will be necessary for optimal performance of this gene delivery vehicle.

References

Acland, G. M., Aguirre, G. D., Ray, J., Zhang, Q., Aleman, T. S., Cideciyan, A. V., Pearce-Kelling, S. E., Anand, V., Zeng, Y., Maguire, A. M., Jacobson, S. G., Hauswirth, W. W. and Bennett, J. (2001). Gene therapy restores vision in a canine model of childhood blindness. Nat. Genet. *28*, 92–95.

Afione, S. A., Conrad, C. K., Kearns, W. G., Chunduru, S., Adams, R., Reynolds, T. C., Guggino, W. B., Cutting, G. R., Carter, B. J. and Flotte, T. R. (1996). In vivo model of adeno-associated virus vector persistence and rescue. J. Virol. *70*, 3235–3241.

Afione, S. A., Wang, J., Walsh, S., Guggino, W. B. and Flotte, T. R. (1999). Delayed expression of adeno-associated virus vector DNA. Intervirology *473*.

Beaton, A., Palumbo, P. and Berns, K. I. (1989). Expression from the adeno-associated virus p5 and p19 promoters is negatively regulated in trans by the rep protein. J. Virol. *63*, 4450–4454.

Berns, K. I. (1990). Parvovirus replication. Microbiol. Rev. *54*, 316–329.

Berns, K. I. and Linden, R. M. (1995). The cryptic life style of adeno-associated virus. Bioessays *17*, 237–245.

Clark, K. R., Sferra, T. J. and Johnson, P. R. (1997). Recombinant adeno-associated viral vectors mediate long-term transgene expression in muscle. Hum. Gene Ther. *8*, 659–669.

Conrad, C. K., Allen, S. S., Afione, S. A., Reynolds, T. C., Beck, S. E., Fee-Maki, M., Barrazza-Ortiz, X., Adams, R., Askin, F. B., Carter, B. J., Guggino, W. B. and Flotte, T. R. (1996). Safety of single-dose administration of an adeno-associated virus (AAV)-CFTR vector in the primate lung. Gene Ther. *3*, 658–668.

Donsante, A., Vogler, C., Muzyczka, N., Crawford, J., Barker, J., Flotte, T. and Sands, M. (2001). Observed incidence of tumorigenesis in long-term rodent studies of rAAV vectors: Are recombinant AAV-based vectors inherently safe? Gene Ther. *8*, 1343–1346.

Duan, D., Yue, Y., Yan, Z. and Engelhardt, J. F. (2000). A new dual-vector approach to enhance recombinant adeno-associated virus-mediated gene expression through intermolecular cis activation. Nat. Med. *6*, 559–598.

During, M. J., Samulski, R. J., Elsworth, J. D., Kaplitt, M. G., Leone, P., Xiao, X., Li, J., Freese, A., Taylor, J. R., Roth, R. H., Sladek, J. R., Jr., O' Malley, K. L. and Redmond, D. E., Jr. (1998). In vivo expression of therapeutic human genes for dopamine production in the caudates of MPTP-treated monkeys using an AAV vector. Gene Ther. *5*, 820–827.

Fisher, K. J., Gao, G. P., Weitzman, M. D., De Matteo, R., Burda, J. F. and Wilson, J. M. (1996). Transduction with recombinant adeno-associated virus for gene therapy is limited by leading-strand synthesis. J. Virol. *70*, 520–532.

Flannery, J. G., Zolotukhin, S., Vaquero, M. I., LaVail, M. M., Muzyczka, N. and Hauswirth, W. W. (1997). Efficient photoreceptor-targeted gene expression in vivo by recombinant adeno-associated virus. Proc. Natl. Acad. Sci. USA *94*, 6916–6921.

Flotte, T. R., Afione, S. A., Conrad, C., McGrath, S. A., Solow, R., Oka, H., Zeitlin, P. L., Guggino, W. B. and Carter, B. J. (1993). Stable in vivo expression of the cystic fibrosis transmembrane conductance regulator with an adeno-associated virus vector. Proc. Natl. Acad. Sci. USA 90, 10613–10617.
Flotte, T., Carter, B., Conrad, C., Guggino, W., Reynolds, T., Rosenstein, B., Taylor, G., Walden, S. and Wetzel, R. (1996). A phase I study of an adeno-associated virus-CFTR gene vector in adult CF patients with mild lung disease. Hum. Gene Ther. 7, 1145–1159.
Hermonat, P. L. and Muzyczka, N. (1984). Use of adeno-associated virus as a mammalian DNA cloning vector: Transduction of neomycin resistance into mammalian tissue culture cells. Proc. Natl. Acad. Sci. USA 81, 6466–6470.
Hermonat, P. L., Labow, M. A., Wright, R., Berns, K. I. and Muzyczka, N. (1984). Genetics of adeno-associated virus: Isolation and preliminary characterization of adeno-associated virus type 2 mutants. J. Virol. 51, 329–339.
Herzog, R. W., Hagstrom, J. N., Kung, S. H., Tai, S. J., Wilson, J. M., Fisher, K. J. and High, K. A. (1997). Stable gene transfer and expression of human blood coagulation factor IX after intramuscular injection of recombinant adeno-associated virus. Proc. Natl. Acad. Sci. USA 94, 5804–5809.
Im, D. S. and Muzyczka, N. (1990). The AAV origin binding protein Rep68 is an ATP-dependent site-specific endonuclease with DNA helicase activity. Cell 61, 447–457.
Kaplitt, M. G., Leone, P., Samulski, R. J., Xiao, X., Pfaff, D. W., O' Malley, K. L. and During, M. J. (1994). Long-term gene expression and phenotypic correction using adeno-associated virus vectors in the mammalian brain. Nat. Genet. 8, 148–154.
Kay, M. A., Manno, C. S., Ragni, M. V., Larson, P. J., Couto, L. B., McClelland, A., Glader, B., Chew, A. J., Tai, S. J., Herzog, R. W., Arruda, V., Johnson, F., Scallan, C., Skarsgard, E., Flake, A. W. and High, K. A. (2000). Evidence for gene transfer and expression of factor IX in haemophilia B patients treated with an AAV vector [see comments]. Nat. Genet. 24, 257–261.
Kearns, W. G., Afione, S. A., Fulmer, S. B., Pang, M. C., Erikson, D., Egan, M., Landrum, M. J., Flotte, T. R. and Cutting, G. R. (1996). Recombinant adeno-associated virus (AAV-CFTR) vectors do not integrate in a site-specific fashion in an immortalized epithelial cell line. Gene Ther. 3, 748–755.
Kessler, P. D., Podsakoff, G. M., Chen, X., McQuiston, S. A., Colosi, P. C., Matelis, L. A., Kurtzman, G. J. and Byrne, B. J. (1996). Gene delivery to skeletal muscle results in sustained expression and systemic delivery of a therapeutic protein. Proc. Natl. Acad. Sci. USA 93, 14082–14087.

Klein, R. L., Meyer, E. M., Peel, A. L., Zolotukhin, S., Meyers, C., Muzyczka, N. and King, M. A. (1998). Neuron-specific transduction in the rat septohippocampal or nigrostriatal pathway by recombinant adeno-associated virus vectors. Exp. Neurol. *150*, 183–194.

Kotin, R. M., Menninger, J. C., Ward, D. C. and Berns, K. I. (1991). Mapping and direct visualization of a region-specific viral DNA integration site on chromosome 19q13-qter. Genomics *10*, 831–834.

Kotin, R. M., Linden, R. M. and Berns, K. I. (1992). Characterization of a preferred site on human chromosome 19q for integration of adeno-associated virus DNA by non-homologous recombination. EMBO J. *11*, 5071–5078.

Kotin, R. M., Siniscalco, M., Samulski, R. J., Zhu, X. D., Hunter, L., Laughlin, C. A., McLaughlin, S., Muzyczka, N., Rocchi, M. and Berns, K. I. (1990). Site-specific integration by adeno-associated virus. Proc. Natl. Acad. Sci. USA *87*, 2211–2215.

Laipis, P., Reyes, L., Embury, J., Alexander, J., Hurt, C., Wein, D., Song, S., Berns, K. I., Chase, D. H. and Flotte, T. (2001). Long-term reduction of serum phenylalanine levels in a mouse model of PKU by rAAV-mediated gene therapy Presented at the 2001 American Society of Gene Therapy.

Mendelson, E., Trempe, J. P. and Carter, B. J. (1986). Identification of the trans-acting Rep proteins of adeno-associated virus by antibodies to a synthetic oligopeptide. J. Virol. *60*, 823–832.

Miao, C. H., Nakai, H., Thompson, A. R., Storm, T. A., Chiu, W., Snyder, R. O. and Kay, M. A. (2000). Nonrandom transduction of recombinant adeno-associated virus vectors in mouse hepatocytes in vivo: Cell cycling does not influence hepatocyte transduction. J. Virol. *74*, 3793–3803.

Miao, C. H., Snyder, R. O., Schowalter, D. B., Patijn, G. A., Donahue, B., Winther, B. and Kay, M. A. (1998). The kinetics of rAAV integration in the liver [letter]. Nat. Genet. *19*, 13–15.

Murphy, J. E., Zhou, S., Giese, K., Williams, L. T., Escobedo, J. A. and Dwarki, V. J. (1997). Long-term correction of obesity and diabetes in genetically obese mice by a single intramuscular injection of recombinant adeno-associated virus encoding mouse leptin. Proc. Natl. Acad. Sci. USA *94*, 13921–13926.

Nakai, H., Yant, S. R., Storm, T. A., Fuess, S., Meuse, L. and Kay, M. A. (2001). Extrachromosomal recombinant adeno-associated virus vector genomes are primarily responsible for stable liver transduction in vivo. J. Virol. *75*, 6969–6976.

Nakai, H., Storm, T. A. and Kay, M. A. (2000). Recruitment of single-stranded recombinant adeno-associated virus vector genomes and intermolecular recombination are responsible for stable transduction of liver in vivo. J. Virol. *74*, 9451–9463.

Peel, A. L., Zolotukhin, S., Schrimsher, G. W., Muzyczka, N. and Reier, P. J. (1997). Efficient transduction of green fluorescent protein in spinal cord neurons using adeno-associated virus vectors containing cell type-specific promoters. Gene Ther. *4*, 16–24.

Pereira, D. J., McCarty, D. M. and Muzyczka, N. (1997). The adeno-associated virus (AAV) Rep protein acts as both a repressor and an activator to regulate AAV transcription during a productive infection. J. Virol. *71*, 1079–1088.

Shi, Y., Seto, E., Chang, L. S. and Shenk, T. (1991). Transcriptional repression by YY1, a human GLI-Kruppel-related protein, and relief of repression by adenovirus E1A protein. Cell *67*, 377–388.

Snyder, R. O., Miao, C. H., Patijn, G. A., Spratt, S. K., Danos, O., Nagy, D., Gown, A. M., Winther, B., Meuse, L., Cohen, L. K., Thompson, A. R. and Kay, M. A. (1997). Persistent and therapeutic concentrations of human factor IX in mice after hepatic gene transfer of recombinant AAV vectors. Nat. Genet. *16*, 270–276.

Snyder, R. O., Miao, C., Meuse, L., Tubb, J., Donahue, B. A., Lin, H. F., Stafford, D. W., Patel, S., Thompson, A. R., Nichols, T., Read, M. S., Bellinger, D. A., Brinkhous, K. M. and Kay, M. A. (1999). Correction of hemophilia B in canine and murine models using recombinant adeno-associated viral vectors [see comments]. Nat. Med. *5*, 64–70.

Song, S., Morgan, M., Ellis, T., Poirier, A., Chesnut, K., Wang, J., Brantly, M., Muzyczka, N., Byrne, B. J., Atkinson, M. and Flotte, T. R. (1998). Sustained secretion of human alpha-1-antitrypsin from murine muscle transduced with adeno-associated virus vectors. Proc. Natl. Acad. Sci. USA *95*, 14384–14388.

Song, S., Laipis, P. J., Berns, K. I. and Flotte, T. R. (2001a). Effect of DNA-dependent protein kinase on the molecular fate of the rAAV2 genome in skeletal muscle. Proc. Natl. Acad. Sci. USA *98*, 4084–4088.

Song, S., Laipis, P., Embury, J., Berns, K., Crawford, J. and Flotte, T. R. (2001b). Stable therapeutic serum levels of human alpha-1 antitrypsin (AAT) after portal vein injection of recombinant adeno-associated virus (rAAV) vectors. Gene Ther. *8*, 1299–1306.

Sun, L., Li, J. and Xiao, X. (2000). Overcoming adeno-associated virus vector size limitation through viral DNA heterodimerization. Nat. Med. *6*, 599–602.

Tratschin, J. D., Miller, I. L. and Carter, B. J. (1984). Genetic analysis of adeno-associated virus: Properties of deletion mutants constructed in vitro and evidence for an adeno-associated virus replication function. J. Virol. *51*, 611–619.

Tratschin, J. D., West, M. H., Sandbank, T. and Carter, B. J. (1984). A human parvovirus, adeno-associated virus, as a eucaryotic vector: Transient expression and encapsidation of the procaryotic gene for chloramphenicol acetyltransferase. Mol. Cell. Biol. *4*, 2072–2081.

Trempe, J. P. and Carter, B. J. (1988). Alternate mRNA splicing is required for synthesis of adeno-associated virus VP1 capsid protein. J. Virol. *62*, 3356–3363.

Virella-Lowell, I., Poirier, A., Chesnut, K. A., Brantly, M. and Flotte, T. R. (2000). Inhibition of recombinant adeno-associated virus (rAAV) transduction by bronchial secretions from cystic fibrosis patients. Gene Ther. *7*, 1783–1789.

Wagner, J. A., Messner, A. H., Moran, M. L., Daifuku, R., Kouyama, K., Desch, J. K., Manly, S., Norbash, A. M., Conrad, C. K., Friborg, S., Reynolds, T., Guggino, W. B., Moss, R. B., Carter, B. J., Wine, J. J., Flotte, T. R. and Gardner, P. (1999). Safety and biological efficacy of an adeno-associated virus vector-cystic fibrosis transmembrane conductance regulator (AAV-CFTR) in the cystic fibrosis maxillary sinus. Laryngoscope *109*, 266–274.

Wagner, J. A., Reynolds, T., Moran, M. L., Moss, R. B., Wine, J. J., Flotte, T. R. and Gardner, P. (1998). Efficient and persistent gene transfer of AAV-CFTR in maxillary sinus [letter]. Lancet *351*, 1702–1703.

Xiao, X., Li, J. and Samulski, R. J. (1996). Efficient long-term gene transfer into muscle tissue of immunocompetent mice by adeno-associated virus vector. J. Virol. *70*, 8098–8108.

Xiao, X., Li, J. and Samulski, R. J. (1998). Production of high-titer recombinant adeno-associated virus vectors in the absence of helper adenovirus. J. Virol. *72*, 2224–2232.

Xiao, X., Li, J., Tsao, Y. P., Dressman, D., Hoffman, E. P. and Watchko, J. F. (2000). Full functional rescue of a complete muscle (TA) in dystrophic hamsters by adeno-associated virus vector-directed gene therapy. J. Virol. *74*, 1436–1442.

Yan, Z., Zhang, Y., Duan, D. and Engelhardt, J. F. (2000). From the cover: Trans-splicing vectors expand the utility of adeno-associated virus for gene therapy [see comments]. Proc. Natl. Acad. Sci. USA *97*, 6716–6721.

Zhou, S., Murphy, J. E., Escobedo, J. A. and Dwarki, V. J. (1998). Adeno-associated virus-mediated delivery of erythropoietin leads to sustained elevation of hematocrit in nonhuman primates. Gene Ther. *5*, 665–670.

Zolotukhin, S., Byrne, B. J., Mason, E., Zolotukhin, I., Potter, M., Chesnut, K., Summerford, C., Samulski, R. J. and Muzyczka, N. (1999). Recombinant adeno-associated virus purification using novel methods improves infectious titer and yield. Gene Ther. *6*, 973–985.

Zolotukhin, S., Potter, M., Hauswirth, W. W., Guy, J. and Muzyczka, N. (1996). A "humanized" green fluorescent protein cDNA adapted for high-level expression in mammalian cells. J. Virol. *70*, 4646–4654.

CHAPTER 2

Production of research and clinical-grade recombinant adeno-associated virus vectors

Joyce D. Francis[1,2] and Richard O. Snyder[1,3]

[1]*Powell Gene Therapy Center,*
[2]*Department of Pediatrics,*
[3]*Department of Molecular Genetics and Microbiology, University of Florida, College of Medicine Box 100296, Gainesville, FL 32610–0296, USA*

Recombinant adeno-associated viral (rAAV) vectors delivered to animal models of disease are capable of sustained gene expression at therapeutic levels with low toxicity (see other chapters in this volume). Demonstration of safety, efficiency, and efficacy following gene delivery in pre-clinical studies, and the subsequent transition to the clinic have been facilitated by improvements in rAAV vector production. Experience in Phase 1 and Phase 2 clinical trials for cystic fibrosis (Flotte et al., 1996; Wagner et al., 1998, 2002), hemophilia (Kay et al., 2000; Manno et al., 2003), muscular dystrophy (Stedman et al., 2000), and Canavan's disease (Janson et al., 2002) using AAV serotype 2 vectors have demonstrated safety and gene transfer in humans. Achieving therapeutic levels of gene expression depends upon the vector elements controlling gene expression, the serotype of the rAAV vector, and the tissue target, where doses ranging from 1×10^{11} vector genomes for retinal gene therapy to 1×10^{14} vector genomes for muscle and liver targets are required. Manufacturing methods are being developed to meet the needs of growing clinical trials involving more patients and higher doses, and ultimately, Food and Drug Administration (FDA) licensed rAAV products.

2.1. Adeno-associated virus biology

Adeno-associated viruses (AAV) were discovered as a contaminant of adenovirus stocks in the late 1960s, identifiable by electron microscopy as small (~20 nm diameter) icosahedral, non-enveloped particles containing single-stranded DNA (Atchison et al., 1966). The International Committee on Taxonomy of Viruses (ICTV) currently classifies AAVs as members of the family *Parvoviridae*, genus *Dependovirus*, and includes five primate AAVs: AAV1, AAV2, AAV3, AAV4, and AAV5; as well as five other AAVs: bovine AAV, canine AAV, avian AAV, equine AAV, and ovine AAV. Recently, three primate variants have been informally designated as AAV6 (Rutledge et al., 1998), AAV7, and AAV8 (Gao et al., 2002), and an additional 30 distinct AAV genomes have been rescued from simian chromosomal DNA (Gao et al., 2003).

Gene transfer experiments in animal models have shown that dramatic differences exist in the transduction efficiency and cell specificity of rAAV vectors of different serotypes (Rutledge et al., 1988; Xiao et al., 1999; Chao et al., 2000; Davidson et al., 2000; Halbert et al., 2001; Gao et al., 2002; Rabinowitz et al., 2002). AAV2 is capable of infecting different types of cells from several species. This broad tropism suggested that the cellular receptor is a common cell surface molecule(s). Indeed, it was shown that the cellular receptor for AAV2 has yielded a complex, which is comprised of three components: heparin sulfate proteoglycan (Summerford and Samulski, 1998), the fibroblast growth factor receptor (FGFR1) (Qing et al., 1999), and $\alpha V\beta 5$ integrin (Summerford et al., 1999). Components of the viral uptake pathway are shared between

AAV2 and its helper viruses: like HSV, AAV2 interacts with the cell surface via heparin sulfate, and AAV internalization is mediated by $\alpha V \beta 5$ integrin, which is also required by adenovirus for uptake. Serotypes other than AAV2 exhibit different tropisms and interact with different cell surface molecules. Some of the cell surface receptors have been identified for AAV3 (HSPG; Rabinowitz et al., 2002), AAV4 (2, 3–0-linked sialic acid; Kaludov et al., 2001), and AAV5 (2, 3-N-linked sialic acid; Kaludov et al., 2001; Walters et al., 2001; PDGF-R, Di Pasquale et al., 2003). The structures of AAV2 and AAV4 capsids have been solved (Xie et al., 2002; Kaludov et al., 2003), and the capsid amino acids important for interacting with their receptors have been predicted using these structures together with capsid mutants (Wu et al., 2000).

AAV2 virions contain a single-stranded DNA genome of 4679 bases (Genbank AF043303; Srivastava et al., 1983) and the other serotypes have genomes of nearly the same size (Chiorini et al., 1997, 1999; Rutledge et al., 1998; Xiao et al., 1999; Gao et al., 2002), with both plus and minus strands packaged equally. Molecular clones of AAV are infectious and have facilitated the study of the genetics of the virus (Samulski et al., 1982; Hermonat et al., 1984; Tratschin et al., 1984). The genome contains two major open reading frames (ORFs). The *cap* ORF encodes the viral capsid proteins VP–1, VP–2, and VP–3 transcribed from the P_{40} promoter which assemble into particles with icosahedral symmetry. The *rep* ORF encodes the four non-structural Rep proteins from two promoters. A spliced and an unspliced mRNA species from both the P_5 and P_{19} promoters are translated to produce the Rep proteins: Rep78, Rep68, Rep52, and Rep40. The P_5 Rep proteins (Rep78 and 68) have been shown to possess functions required to replicate the genome (Im and Muzyczka, 1990; Owens et al., 1993; Smith et al., 1997), modulate transcription from AAV and heterologous promoters (Labow et al., 1986; McCarty et al., 1991; Horer et al., 1995), and mediate site-specific integration into the human genome (Weitzman et al., 1994). The P_{19} Rep proteins (Rep52 and 40) are required for the production of single-stranded genomes, insertion of the genome into the capsid, and the modulation of AAV promoters (Chejanovsky and Carter, 1989; Smith and Kotin, 1998; King et al., 2001). The Rep proteins have been shown to interact with cellular proteins involved in gene expression (Pereira and Muzyczka, 1997; Chiorini et al., 1998; Di Pasquale and Stacey, 1998; Weger et al., 1999). The terminal genome sequences are 145 nucleotides for AAV2 which form T-shaped palindromic structures and encode sequences required for packaging, integration, and rescue, and also serve as the origins of DNA replication (McLaughlin et al., 1988).

AAV requires both host cell factors and coinfection with adenoviruses (Ad) or herpesviruses for a productive infection. During this lytic cycle, greater than 10^6 genomes, 10^7 preformed capsids, 10^6 total virions, and 10^4 infectious virions are generated per cell (Carter, 1990). The Ad gene products that play a role in the AAV lytic cycle have been delineated and include E1A, E1B, E2A, E4, and VA (Richardson and Westphal, 1981). The Ad helper functions act throughout the AAV replicative cycle to promote AAV production. During an Ad infection, E1A gene products transcriptionally activate the Ad early promoters for Ad gene expression, and an E1A inducible element is present in the AAV P_5 promoter. The regulation of the P_5 promoter has been shown to be mediated by a host factor YY1, that represses the P_5 promoter in the absence of the Ad E1A protein and activates P_5 in its presence (Shi et al., 1991). Two products from delayed early transcripts, the E1B 55K protein and E4 34K protein, form a complex in infected cells that controls the accumulation of Ad and AAV mRNAs in the cytoplasm (Pilder et al., 1986). The E2A 72K DBP is a multi-functional protein, synthesized both early and late, and AAV requires its translational regulatory activities (Jay et al., 1981) and its ability to stimulate transcription from the P_5 promoter (Chang and Shenk, 1990). The VAI RNA is needed for efficient translation of adenoviral and AAV mRNA (Janik et al., 1989). The herpesviruses can also supply helper functions, and the HSV genes responsible for helping AAV have distinct functions from the Ad genes and include a subset of genes required for HSV DNA replication (Weindler and Heilbronn, 1991).

The most unique feature of AAV is the latent phase of the life cycle. When AAV infects a permissive cell in the absence of helper virus, stable latency is established, without obvious consequences for the host cell (Kotin et al., 1990). The Rep proteins can be cytostatic or cytotoxic, depending on the level of expression, and so their expression is tightly suppressed in latency. The viral genomes in latent infection can assume a number of integrated and episomal forms (McLaughlin et al., 1988; Nakai et al., 2001; Huser et al., 2002; Schnepp et al., 2003), but the predominant form for AAV2 infection in human cells appears to be a tandem head-to-tail integrated form within a region of human chromosome 19, that has been designated AAVS1 (Kotin et al., 1990; Samulski et al., 1991). It has been demonstrated that this integration event, like replication, is dependent upon the Rep78 and Rep68 proteins, which are capable of binding both the AAVS1 and the AAV2 inverted terminal repeat (ITR) in a single complex (Weitzman et al., 1994), and are capable of nicking both sites as well (Urcelay et al., 1995).

2.1.1. Recombinant AAV vectors

Recombinant AAV vectors are constructed by replacing the *rep* and *cap* genes with the transgene of interest while retaining the flanking ITR sequences (Samulski et al., 1989). Construction of rAAV vectors is possible because the AAV inverted terminal repeats (ITRs) supply all of the *cis* acting sequences for vector production and transduction (McLaughlin et al., 1988; Samulski et al., 1989). A transgene can replace the AAV coding sequences because both the *rep* and *cap* gene products can be supplied in *trans* to make infectious rAAV virions (Samulski et al., 1989). Recombinant AAV vector constructs are flanked by AAV ITRs (e.g., AAV2 ITRs comprise the pTR-UF backbone (Zolotukhin et al., 1996; Klein et al., 1998)) and other *cis*-sequences control transgene expression including: constitutive or regulated (exogenously or tissue-specific) promoters, enhancers, WPRE, IRES, polyA, and intron sequences.

2.1.2. rAAV serotype vectors

In general, there are two different approaches for packaging rAAV vectors: "isotype" and "pseudotyped" vectors. The former refers to vectors having inverted terminal repeats (ITRs), Rep proteins, and capsid proteins derived from the same wild-type virus (Hermonat and Muzyczka, 1984; Chiorini et al., 1997, 1999). The latter refers to vectors derived from ITRs and Rep proteins of one serotype virus, and capsid proteins of another, e.g., 2 and 1 (AAV2/1) (Xiao et al., 1999; Halbert et al., 2000; Hildinger et al., 2001; Rabinowitz et al., 2002). The pseudotyping of AAV2-ITR containing vectors is preferred because more experience exists with the safety profile of these ITRs in animal models and humans. The chromosomal integration efficiency and specificity has been investigated for AAV2 ITRs, but little data has been generated thus far with the ITRs of other AAV serotypes.

2.2. rAAV preparation

2.2.1. rAAV vector production

In recent years, there have been significant improvements in production and purification of rAAV vectors. The major improvements in production have included enhanced output of the number of DNase resistant particles (drp) per cell and the emergence of scaleable systems. The most widely utilized rAAV vector production methods require four genetic elements (Hermonat and Muzyczka, 1984; Tratschin et al., 1984): (1) mammalian tissue culture cells, (2) vector sequences containing a transgene flanked by AAV inverted terminal repeats, (3) AAV helper sequences comprising the AAV open reading frames (ORFs), and (4) helper virus (i.e., adenovirus or herpesvirus) genes which act throughout the AAV lytic cycle, including activation of AAV gene expression. Replication and packaging of these vectors is accomplished by providing permissive cells with Rep, Cap, and the adenovirus helper gene products (E1a, E1b, E2a, E4, and VA-RNA), or the required herpesvirus gene products. Originally, helper functions were provided by infecting cells with helper virus prior to cotransfection with a rep/cap-expressing helper and vector plasmids. Several groups independently found that the use of plasmids to express adenovirus (Ad) helper genes in transient transfection results in greater efficiency of rAAV production than infection with Ad virus, perhaps because of enhanced viability of producer cells or the lack of competition with the helper virus for DNA replication machinery (Ferrari et al., 1997; Grimm et al., 1998; Matsushita et al., 1998; Xiao et al., 1998). Supplying the adenoviral genes through infection is a convenient way to induce AAV gene expression, but the Ad that is generated must be physically removed or inactivated (inactivation by itself does not remove the cytotoxic Ad structural proteins). Another interesting finding is that down-regulation of Rep78/68 relative to Rep52/40 and the capsid proteins results in a greater accumulation of single-stranded DNA genomes and packaged vector DNA (Li et al., 1997; Grimm et al., 1998). The incorporation of these improvements into transient transfection production protocols has enhanced yields from about 100–1000 drp per cell to over 10,000 drp per cell.

Other protocols involve cell lines with integrated rep/cap cassettes (Clark et al., 1995; Gao et al., 1998; Liu et al., 1999; Chadeuf et al., 2000; Mathews et al., 2002; Qiao et al., 2002a,b) infected with adenovirus or, alternatively, a recombinant herpesvirus system has been used to provide both helper virus function and rep/cap (Conway et al., 1997, 1999). In a switch away from using mammalian cell and helper virus production systems, rAAV vectors have been made in insect cells where the AAV genes are expressed under the control of insect promoters and the traditional helper virus gene products are not required (Urabe et al., 2002). Stable producer cell lines, and packaging cell lines used in combination with recombinant hybrid AAV-adenoviruses have achieved 10,000–30,000 drp per cell. Hybrid AAV-herpes vectors have achieved outputs that approach the 5×10^5 drp per cell seen with wtAAV. Yields using the baculovirus system approach 5×10^4 drp per Sf9 cell. Overall, these newer methods produce greater vector yields, and reduce or eliminate detectable replication-competent AAV (rcAAV) contamination. The ability to produce sizable stocks of recombinant AAV that are free of contaminating wild-type virus can be challenging. Homologous and illegitimate recombination between vector and AAV helper plasmids may result in the generation of wild-type AAV, albeit at very low levels (Allen et al., 1997; Salvetti et al., 1998; Wang et al., 1998). However, any wild-type AAV that is generated during production is non-pathogenic and will require a helper virus for growth.

2.2.2. rAAV vector purification

In previous purification methods for rAAV, the crude cell lysate was normally partially purified prior to loading on CsCl gradients by the use of batch methods such as precipitation with ammonium sulfate or polyethylene glycol (Snyder et al., 1996). This was then followed by at least two rounds of CsCl centrifugation. However, rAAV appears to interact with cellular and viral proteins in this high ionic strength environment. These complexes fail to display uniform biochemical properties, making it difficult to develop a reproducible purification strategy. This also often leads to poor recovery and carry over of the contaminating proteins even after several rounds of CsCl gradient centrifugation. Furthermore, CsCl often results in vector stocks that have very high particle to infectivity (P:I) ratios, implying that much of the vector has been inactivated. Improvements in downstream purification have generally involved a departure from CsCl density gradients and towards alternative density gradient media or column chromatography methods.

Several features of parvoviral virions make column chromatography a viable option for their purification (Arella et al., 1990). The virions are very small (18–26 nm) and stable to a wide range of temperatures (–80 to 56 °C), pH 3–9, sonication, solvents ($CHCl_3$), detergents, proteases, and nucleases. The virions are also stable during precipitation using $(NH_4)_2SO_4$, PEG, and $CaCl_2$ (Arella et al., 1990; Snyder et al., 1996). Purification using affinity chromatography, based on identified cellular receptors (Summerford and Samulski, 1998; Kaludov et al., 2001), is becoming more common and the more physiological conditions (versus CsCl gradients) result in vector stocks with P:I ratios of <100.

2.3. Protocols

2.3.1. Cell transfection

Cell and virus processing is performed exclusively in biosafety cabinets during open steps. HEK293 cells (ATCC) are cultured in DMEM supplemented with 5% fetal bovine serum and antibiotics (called DMEM-complete). PBS and 0.05% trypsin are used during cell passage. HEK293 cells are split 1:3 the day prior to transfection, so at the time of transfection the cell confluency is ~75–80%. A production run utilizes about 1×10^9 cells seeded in a Cell Factory (Nunc) and CsCl purified plasmid DNA. The $CaPO_4$-precipitate is formed by mixing the helper plasmid, and the rAAV vector plasmid at a 1:1 molar ratio (total of 2.4 mg of DNA) in a total volume of 50 ml of 0.25 M $CaCl_2$, followed by the addition of 50 ml of 2× HBS pH 7.05 to the DNA/$CaCl_2$. The mixture is incubated for 1–2 min at room temperature, at which time the formation of precipitate is stopped by diluting the mixture into 1100 ml of pre-warmed DMEM-complete. The conditioned culture media is removed from the cells and the fresh precipitate-containing media is added immediately. Cells are incubated at

37 °C, 5% CO_2 for 60 h and the $CaPO_4$ precipitate is allowed to remain on the cells during this incubation period. At the end of the incubation the culture media is discarded, cells are washed with PBS, and harvested using PBS containing 5 mM EDTA. The collected cells are centrifuged at $1000 \times g$ for 10 min, and stored at -20 °C until purified.

2.4. Small-scale rAAV 1, 2, and 5 vector purification

An efficient and reproducible protocol based on partial purification of an initial freeze/thaw lysate by iodixanol gradient fractionation, followed by chromatography on heparin-sepharose for the purification and concentration of rAAV2 vectors has been reported (Zolotukhin et al., 1999). rAAV1, 2 and 5 vectors are purified by iodixanol gradient centrifugation and anion-exchange (Q-Sepharose) chromatography (Zolotukhin et al., 2002). The resulting vector stocks are 99% pure with titers of 1×10^{12} to 1×10^{13} drp/ml.

2.4.1. Cell lysate and iodixanol gradients

Cells are lysed by three freeze/thaw cycles between dry ice–ethanol and 37 °C water baths. Benzonase (Sigma) is then added to the cell lysate (50 U/ml final concentration) and incubated for 30 min at 37 °C. The crude lysate is clarified by centrifugation at $4000 \times g$ for 20 min and the vector-containing supernatant is divided between four iodixanol gradients.

Discontinuous iodixanol step gradients are formed in quick seal tubes (25 × 89 mm, Beckman) by underlaying and displacing the less dense cell lysate (15 ml) with iodixanol (5,5′[(2-hydroxy-1-3-propanediyl)-bis(acetylamino)] bis [N,N'-bis(2,3dihydroxypropyl-2,4,6-triiodo-1,3-benzenecarboxamide]) prepared using a 60% (w/v) sterile solution of OptiPrep (Nycomed) and PBS-MK buffer (1 × PBS containing 1 mM $MgCl_2$ and 2.5 mM KCl). Therefore, each gradient consists of (from the bottom): 5 ml 60%, 5 ml 40%, 6 ml 25%, and 9 ml of 15% iodixanol; the 15% density step also contains 1 M NaCl. Tubes are sealed and centrifuged in a Type 70 Ti rotor at 69,000 rpm (350,000 g) for 1 h at 18 °C. Approximately 5 ml of the 60–40% step interface is aspirated after side-puncturing each tube with a syringe equipped with an 18-gauge needle. The iodixanol band from each of the four gradients is combined; this can be stored refrigerated or frozen until column chromatography is performed.

2.4.2. rAAV column chromatography

The iodixanol gradient fraction is further purified and concentrated by column chromatography. For AAV2 vectors, a 3 ml heparin agarose Type I column (Sigma) is equilibrated with 10 ml of PBS-MK buffer, then 10 ml of PBS-MK/1 M NaCl, followed by 20 ml of PBS-MK buffer. The vector-containing iodixanol fraction (20 ml) is loaded onto the column by gravity flow. The column is washed with 20 ml of PBS-MK buffer and eluted in 15 ml of PBS-MK/1 M NaCl. Alternatively, the AAV2 vectors are purified using a 1 or 5 ml HiTrap Heparin column (Pharmacia) on an ATKA FPLC system (Pharmacia) run at 1 column volume (CV) per minute and eluted with PBS-MK/0.5 M NaCl.

For rAAV1 and 5 vectors, a 5 ml HiTrap Q column (Pharmacia) is equilibrated at 5 ml/min with 5 column volumes (25 ml) of Buffer A (20 mM Tris, 15 mM NaCl, pH 8.5), then 25 ml Buffer B (20 mM Tris, 500 mM NaCl, pH 8.5), followed by 25 ml of Buffer A using a Pharmacia ATKA FPLC system. The 20 ml vector-containing iodixanol fraction is diluted 1:1 with Buffer A and applied to the column at a flow rate of 3–5 ml/min. After loading the sample, the column is washed with 10 column volumes (50 ml) of Buffer A. The vector is eluted with Buffer B and fractions are collected.

2.4.3. Vector concentration

If necessary, the vector can be concentrated, and desalted or buffer exchanged in a Biomax 100 K concentrator (Millipore) by three cycles of centrifugation. In each cycle the virus is concentrated to 1 ml following the addition of 10 ml of Lactated Ringer's or 1× PBS. The virus is stored at –80 °C in lactated Ringer's or 1× PBS.

2.5. Large-scale rAAV purification

Large-scale production of rAAV vectors requires a fully scaleable purification method to accompany the upstream production process. The purification scheme described above for small-scale rAAV preparations involves a single iodixanol gradient step followed by chromatography. Although the chromatography step is easily amenable to scale-up, iodixanol gradient centrifugation is not readily scaled.

2.5.1. Large-scale purification of AAV2

Large-scale purification by column chromatography is made possible by two changes in the preparations of crude lysates. The first is the use of a microfluidizer to lyse cells and form a fine suspension of cellular debris. The second came from an observation made by Johnson and his collaborators (Clark et al., 1999). They introduced the use of deoxycholate in crude viral suspensions, which appears to reduce viral aggregation. An example of this kind of purification is shown in Figs 2.2B and 2.3. A crude lysate is prepared from ten cell factories (1×10^{10} cells) by resuspending the thawed cells in lysis buffer (20 mM Tris pH 8, 15 mM NaCl, 0.5% deoxycholate containing 50 U/ml benzonase), followed by microfluidization. The lysate is loaded immediately onto a 150 ml Streamline Heparin (Pharmacia) column at the rate of 20 ml/min using an AKTA-FPLC (Pharmacia). The column is washed with 4 Column Volumes

(CV) of lysis buffer and then 5 CV of PBS. The UV absorbance is monitored at 280 nm while the virus is eluted with PBS containing 0.5 M NaCl. The peak is collected (approximately 90 ml), Fig. 2.3A. The NaCl concentration of the peak is adjusted to 1 M and subjected to hydrophobic interaction chromatography on Phenyl sepharose (5 ml column, Pharmacia); the flow-through is collected, Fig. 2.3B. The flow-through is diluted to ~150 mM NaCl by diluting with water. The vector is then loaded on a 5 ml SP sepharose column (Pharmacia) at 5 ml/min, washed with 10 column volumes of PBS and eluted with PBS containing NaCl (~0.5 M), Fig. 2.3C. Absorbance is monitored at 280 nm, the peak is collected and the presence of virus is confirmed in peak fractions by infectious titer, vector genome titer, and silver staining of SDS-PAGE gels. In a typical large-scale run, the vector genome titers (DNase resistant particles) average $2-3 \times 10^{13}$ ml^{-1} in a total of 10 ml. The particle to infectivity ratios are less than 200 and the purity as judged by silver stain gel analysis is >95% pure (Fig. 2.4A).

2.6. rAAV vector characterization

Having reliable, specific, and sensitive assays is crucial during development of production and purification methods, and to show lot consistency. Once made and purified, the rAAV vector stocks are characterized using a variety of techniques to ensure they are safe, pure, potent, and stable.

2.6.1. Assay for protein purity of the rAAV stock

Stocks are analyzed by silver or Coomassie blue staining of capsid proteins separated on SDS polyacrylamide gels. As shown in Fig. 2.4A, the three capsid proteins (VP1, 2, and 3) are visible in the correct stoichiometry of 1:1:10, and are free of non-AAV proteins (>95% pure).

2.6.2. Assay for infectious rAAV

The infectious center assay (ICA) assay is used to determine the infectious titer of rAAV. The assay measures the ability of the virus to infect cells, unpackage, and replicate. Its major feature is that it gives an accurate measurement of infectious virus regardless of the transgene or promoter used. The method, however, provides no information as to the functional status of the expression cassette delivered by the viral particle. In this assay, rep/cap-expressing C12 cells, (Clark et al., 1996) are infected with serial dilutions of rAAV and superinfected with adenovirus at a MOI of 5–20 for 40 h. Cells that have been infected by rAAV are then complemented for DNA replication and amplify the rAAV genome to several thousand copies per cell. When the cells are subsequently trapped on nylon filters, lysed by treatment with NaOH, and probed for transgene DNA, only those cells that have been productively infected with rAAV produce a spot on film (Fig. 2.4B). The assay is accurate in the range of 20–200 spots (or infectious centers) per filter. The switch to C12 cells also allows the elimination of wtAAV to supply *rep* and *cap* in the ICA (Snyder et al., 1996). This removes a potential source of contamination from the rAAV stocks.

2.6.3. Assay for contaminating replication-competent AAV

The infectious center assay can also be used to measure the level of rcAAV contamination in an rAAV preparation. In this case, dilutions of the rAAV preparation are used to infect HEK293 cells in the presence of adenovirus. Since no additional wild-type AAV is added, only particles in the rAAV preparation that have packaged a wild-type or pseudo-wtAAV genome are capable of replication and produce a signal following probing with the AAV *rep* gene.

2.6.4. Dot blot assay to determine the titer of rAAV physical particles and the particle to infectivity ratio

The dot blot assay is used to determine the titer of rAAV genome-containing virions (Fig. 2.4C), however, it does not indicate virus infectivity or functionality of the expression cassette. DNA external to the virion is digested with nuclease, the genome is liberated from the virion by digestion with proteinase K, and the rAAV DNA is extracted and fixed to a nylon membrane along with a standard dilution series of the plasmid DNA that was packaged. The blots are probed for the transgene and compared to known amounts of plasmid standard DNA to determine titer in units of drp. This data together with the infectious titer can be used to determine the particle to infectivity ratio for a vector preparation. Ratios at or below 200:1 are routinely achieved and higher ratios indicate viral inactivation that is likely to result in poor transduction efficiency.

2.6.5. ELISA for determination of total AAV particles

The AAV2 capsid ELISA assay is used quantitate the total (infectious, non-infectious, and empty) viral particles. Grimm et al. (1999) reported that some preparations of rAAV may have a significant amount of empty particles. The A20 antibody recognizes assembled empty or full AAV2 capsids, but not disrupted or partially assembled capsids. This assay can determine: (1) what fraction of the rAAV particles are empty and (2) whether current purification schemes are effective at removing empty particles.

2.6.6. Assay for infectious helper virus contamination of rAAV

Transfection methods that use a mini-Ad plasmid DNA system (Grimm et al., 1998; Grimm, 2002; Zolotukhin et al., 2002) to supply Ad helper functions and AAV genes are incapable of making infectious Ad. However, if a helper virus is utilized (Clark et al., 1995; Conway et al., 1999; Gao et al., 1998; Sun et al., 2003), assays are available to monitor their presence. The first is the standard plaque assay on 293 cells for Ad or on Vero cells for HSV. The second is a $TCID_{50}$ assay where cytopathic effect (CPE) is scored after serial dilution of the rAAV vector and infection of suitable complementing cells.

2.6.7. Assay for transgene expression

The assay for transgene expression involves the transduction of tissue culture cells or animals, and testing for the presence of the transgene protein product. Reagents (antibodies, enzymatic substrates, nucleic acid probes) specific for the transgene product should be made in large quantities, so the assay can be repeated several times.

2.6.8. Stability program

Demonstration of product stability at the proper storage temperature in its storage container and final formulation is necessary. The program needs to demonstrate genetic and physiochemical stability, including sterility (package integrity) and potency (infectious titer) over time. The testing interval as well as duration needs to be determined prior to initiating the stability study; in some cases the study can run concomitantly with the clinical trial.

2.7. Safety testing

Adventitious agents (in vitro and in vivo). These assays are designed to detect the presence of infectious viral agents of human or animal origin (serum, trypsin).

Mycoplasma. The test for the presence of mycoplasma relies on expansion of cells in antibiotic-free conditions and detection of the organism using dye or PCR, as well as growth of mycoplasma on appropriate agar media.

Endotoxin. Endotoxin can be detected using the LAL or rabbit pyrogen assays.

Sterility. This assay determines the absence of bacterial or fungal organisms.

General safety. This assay is designed to determine if there are toxins (chemical or biological) that would induce an acute toxicity in animals.

2.8. Pre-clinical regulatory compliance activities

Regulatory compliance is usually not of major concern during the production of AAV vectors for use in the research laboratory. Early studies may be performed with vector produced in any manner that yields the type and amount of vector required. It is critical, however, that fairly detailed research notes be recorded contemporaneously. This information is important to provide strong support for future development, production and testing efforts, as well as adequate background information to pursue clinical research. The source of each component used or designed should be clearly defined. When the researcher believes that the vector is a promising candidate for therapeutic use, additional attention must be paid to compliance with US and, possibly, international regulations related to the development of a vector for use in clinical trials.

Like other potential biopharmaceutical products, FDA, Center for Biologics Evaluation and Research (CBER) regulates the production, testing and use of AAV vectors in human clinical trials, in the US. Many regulatory documents are available to help guide the researcher. These documents are available via the Internet on http://www.fda.gov.cber/guidelines.htm. Review of these documents prior to significant work towards production of vector for clinical trial use is necessary to plan and implement an appropriate product development program.

Safety is the first concern of regulatory authorities when a potential gene therapy product is to be evaluated under an IND in humans. Non-clinical laboratory studies to support IND applications for potential products regulated by the FDA, should be performed in compliance with Good Laboratory Practices (GLPs) regulations, 21 CFR Part 58. Non-clinical studies are defined in these regulations as in vivo or in vitro experiments in which the vector (test article) is studied prospectively in test systems (animal models, cell culture, etc.) under laboratory conditions to determine the vector's safety. Safety studies for AAV vectors include acute and long-term toxicology, and biodistribution of vector in appropriate animal models. Other safety studies may be required. Research studies that are considered characterization of the vector, such as expression levels, efficacy, etc., or studies to determine the vector's physical or chemical characteristics do not need to be performed in a manner compliant with GLP regulations. GLP compliance ensures the quality and acceptability of the safety data generated in these studies.

Characterization of the vector batch used in non-clinical safety studies is also performed in compliance with GLPs. Care should be taken to ensure that adequate and appropriate processes are used to produce the batch(es) required, and the production process should be similar to that proposed for production of the vector for human clinical trials, or at least performed in such a way that adequate documentation of the production methods is available for comparative purposes. The GLP regulations state that the identity, purity, and composition of the vector batch (test article) used in a safety study must be known and documented. In addition, the stability of the vector preparation in the specific container used for the study must also be known prior to initiation of the study or acquired concomitantly with the study itself. These test article characterization experiments require that the researcher produce additional amounts of vector identical to the material used in

TABLE 2.1

Characterization of AAV vector batches used in animal safety studies

Sterility	Potency–infectious titer
Mycoplasma	Strength–particle/genome titer
Endotoxin	Protein purity
rcAAV	

the safety studies. AAV vector batches can be characterized for use in safety studies by performing the tests listed in Table 2.1.

For stability of the vector in its storage container, potency and sterility testing should be performed repeatedly during the course of the non-clinical study, particularly during the dosing phase. It is important to document that no change in potency of the vector preparation occurred during the study due to container, dosing formulation, or storage condition. The characterization data should be generated in a GLP compliant laboratory to ensure the quality of the test article batch used in a safety study and hence the reliability and appropriateness of the safety data.

Academic researchers may find it difficult to establish testing facilities compliant with GLPs within their own institutions. Crucial to compliance is the formation of a Quality Assurance Unit (QAU) separate from and independent of the personnel engaged in the direction and/or conduct of the safety study. The QAU is responsible for monitoring the study to assure that the study is performed in compliance with GLPs. The researcher may turn to a National Gene Vector Laboratory or commercial GLP testing facilities for performance of safety testing. Conducting safety studies in compliance with GLPs is the responsibility of the institution submitting the research permit application (IND), even if testing is performed elsewhere. Contract testing facilities should be audited by the researcher prior to initiating the safety study. Review of the facility, operations, standard operating procedures, quality assurance unit, and the technical appropriateness of their personnel is recommended. Visiting the testing facility during the course of the

TABLE 2.2
Pre-clinical testing facility review

- Quality Assurance Unit—independence, master schedule, audits, and inspections
- Document Control Systems—SOPs, protocols, protocol amendments, final reports
- Receipt, storage and handling of test and control article
- Animal care and handling SOPs
- Study director qualifications and responsibilities
- Training documentation for study personnel including annual GLP training
- Equipment records—daily monitoring and routine maintenance and calibration
- Reagents and solutions controls
- Prevention of cross contamination
- Facilities – size, appropriate separation of activities and studies
- Storage and retrieval of records and data
- Retention of records

GLP study ensures that the testing facility is operating according to their written procedures and that appropriate data is being collected and documented. GLP compliance is founded in documentation, so ensuring that systems are in place to archive all appropriate information applicable to the study is important. Since the timeline for submitting an IND to licensing a product is lengthy and the inspection at the testing facility of the study data by a regulatory authority may occur years after it is performed, ensuring compliance during the study reduces the likelihood of having to repeat the study. Table 2.2 provides a list of areas that should be reviewed with the testing facility at their location.

2.9. Manufacture of clinical-grade rAAV vectors

The transition of gene transfer experiments from laboratory to clinic requires the incorporation of control systems for manufacturing, quality, and the clinic. Clinical-grade rAAV vectors are manufactured according to current Good Manufacturing Practices (cGMPs) as outlined in the Code of Federal Regulations (21CFR) or European Commission Directives (91/356/EEC), for drug applications submitted to the Food and Drug Administration or European Medicines Evaluation Agency, respectively. To meet these requirements, an appropriate process needs to be developed.

2.9.1. Technology transfer

The manufacture of clinical-grade rAAV vectors requires the development of a process that will provide an adequate number of safe doses of vector for the trial itself as well as for release testing and for the examination of the vector product's stability in the final formulation over time. Most research-level processes would not meet this requirement. During the development of the clinical-grade process, it is important to consider many factors including scale, vector yield, product purity, process impurities, sampling procedures, and final product formulation. The use of closed systems that can be cleaned and sanitized adds to the safety and quality of the final vector product. Clearly, a process giving the highest yield, purity, and potency is most desirable. Process impurities such as host cell DNA and process reagent residuals should be minimized. The process steps should lend themselves to appropriate in-process sampling points (Figs 2.1–2.4) to assess the reproducibility and success of the process, and the needed samples to meet release requirements. The formulation of the final product must be examined to provide not only the appropriate dose in an acceptable excipient, but also a product that is stable when stored and shipped to distant clinical sites. As the product matures through the development cycle, i.e., Phase 1, Phase 2 and finally Phase 3 clinical trials, process development continues to meet the increased dose requirements of later trials and the tightening of manufacturing controls.

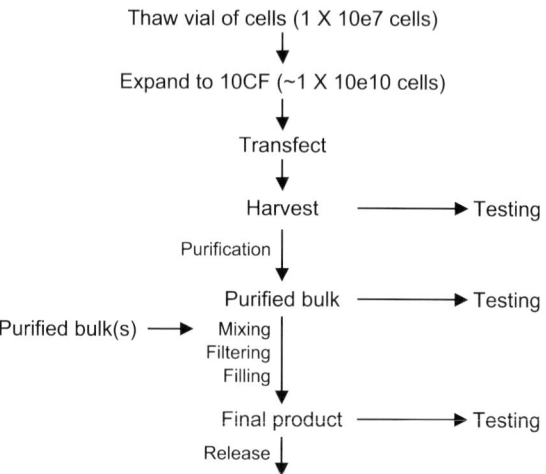

Fig. 2.1. Overview of clinical rAAV manufacturing. See text for details.

2.9.2. Quality control

To support the process development cycle and the manufacture of clinical-grade vector, a quality control (QC) function is needed. The quality of the vector stocks needs to be analyzed by several criteria in order to develop a manufacturing process and compare results from animals transduced with different vector lots. The level of protein contaminants and the proper ratio of the AAV capsid proteins can be assessed on stained denaturing protein gels, and by immunoblotting using antibodies to possible contaminants (such as adenovirus proteins and bovine serum albumin) and AAV capsid proteins. Levels of infectious helper virus (Ad or HSV) can be evaluated using plaque assays, immunofluorescence staining for viral proteins, or CPE assays. Contaminating cellular, helperviral, and plasmid DNA can be assessed by PCR. The contamination of vector stocks with plasmid DNA could greatly impact the efficacy of the gene therapy, as unmethylated CpG dinucleotides present in plasmid DNA produced in *E. coli* can elicit a potent

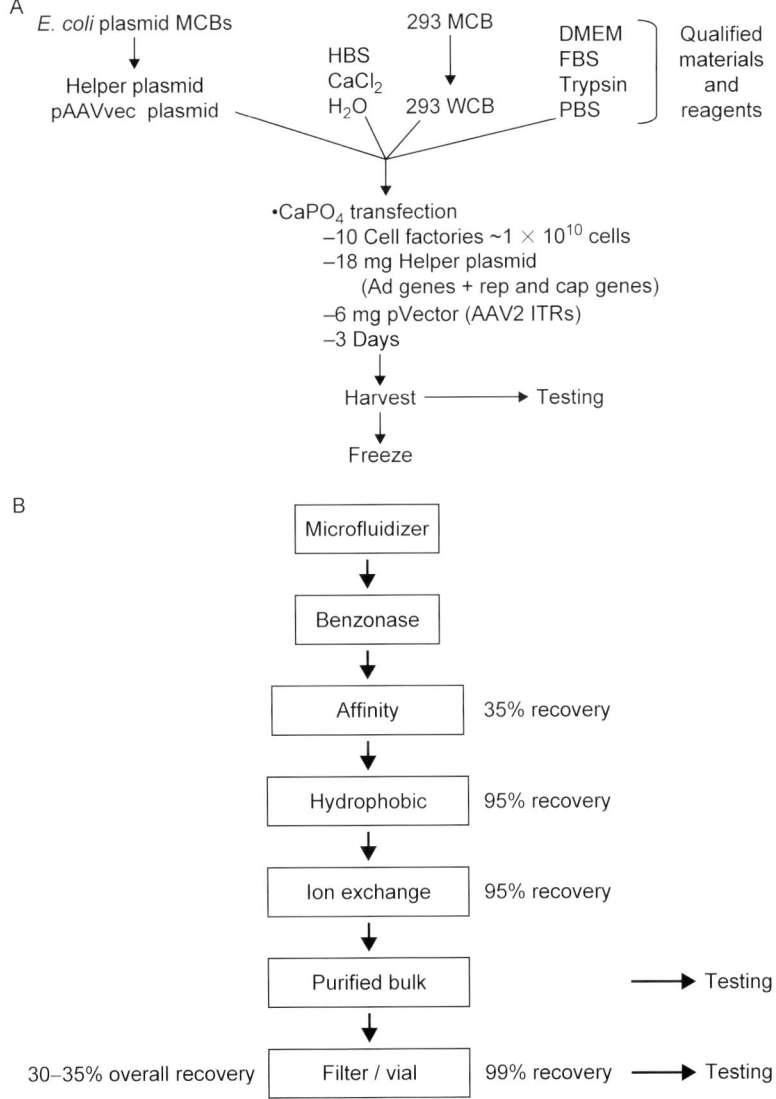

Fig. 2.2. Clinical manufacturing processes. (A) Upstream processes include raw materials qualification, cell expansion, transfection, and harvesting. (B) Downstream processes include cell lysis and clarification, chromatography, and filtering, vialing, and storage. Recoveries are indicated for each step.

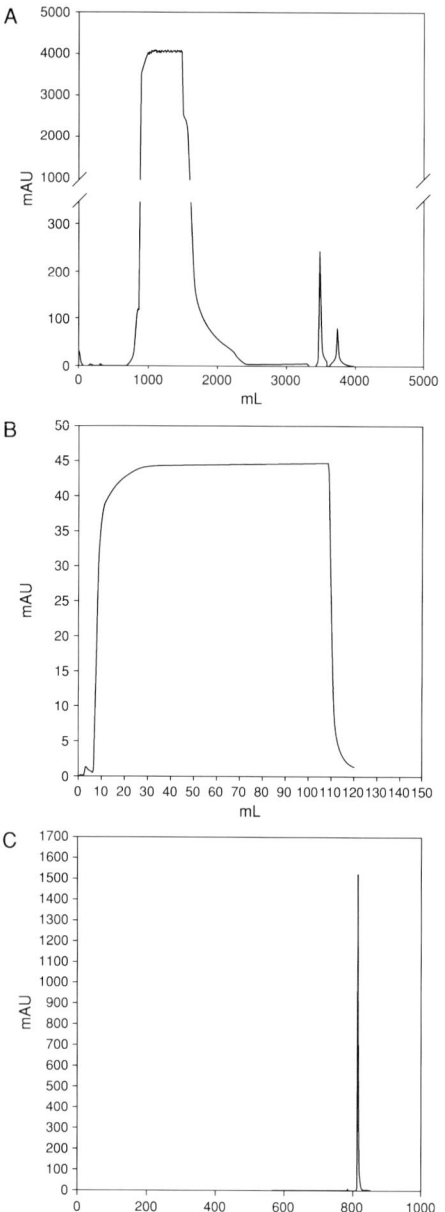

Fig. 2.3. FPLC purification of rAAV. Ten Nunc cell factories of rAAV-UF11

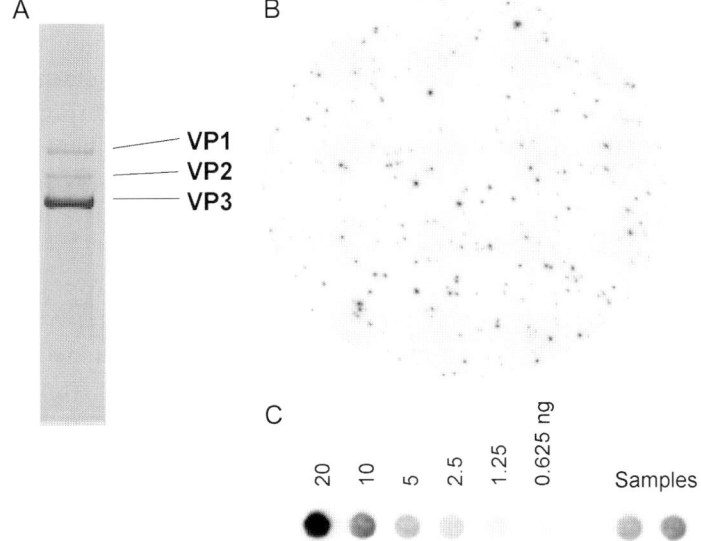

Fig. 2.4. Vector characterization. (A). SDS acrylamide gel electrophoresis of a rAAV-UF-11 vector. Approximately 2.5×10^{11} drp were loaded on a 10% SDS-polyacrylamide gel. The three capsid protein bands in the correct stoichiometry are visualized by silver staining. (B) Infectious center assay of a rAAV-CB-hAAT stock when probed for the transgene; each spot represents a vector infected cell. (C) Dot blot analysis of the rAAV stock to determine the titer of DNase-resistant particles of rAAV. Amounts (ng) refer to a serial dilution of the packaged rAAV vector plasmid used to construct the standard curve. The stock has a calculated titer of 4.3×10^{13} drp/ml.

immune response (Krieg et al., 1998). The degree of rcAAV generated through homologous or non-homologous mechanisms can be determined using the replication center assay, PCR, or Southern blot analyses.

---

were microfluidized in the presence of deoxycholate and benzonase, and loaded on the first of three FPLC columns. The first column is a streamline (Pharmacia) heparin column. The second column is phenyl sepharose and the third column is SP sepharose. Shown are tracings for A_{280}, see Section 2.3 for details.

During clinical-grade vector production, quality control personnel have the responsibility to: (1) monitor the manufacturing facility to determine if it is within the specifications for cleanliness (microbial and particulates), (2) monitor manufacturing personnel hygiene, (3) evaluate raw materials to ensure adherence to specifications, (4) evaluate the manufacturing process by testing intermediates and, (5) characterize the product (identity, purity, potency, stability [genetic and physiochemical]) and perform product safety testing (sterility, endotoxin, mycoplasma, adventitious agents). The level of endotoxin present in the vector preparations needs to be low because administration of contaminated preparations can induce sepsis. Sterility of the preparation should be maintained and the formulation should be physiological. Total and infectious particle titers need to be determined to evaluate dosing and the total particle:infectious particle ratio. The assays for the infectious particle number should incorporate treatments to facilitate the most efficient transduction (Ferrari et al., 1996; Mah et al., 1998). Finally, several other quality criteria have been established for the clinical use of gene transfer vectors (CBER, 1998).

2.10. Clinical manufacturing regulatory compliance activities

As the suitability of the vector for clinical trial use becomes apparent during research and pre-clinical safety evaluations, manufacturing controls and quality systems should be developed. Manufacturing controls include: (1) adequate cleanroom facilities and equipment that are calibrated, validated, and properly cleaned and maintained, (2) personnel who have relevant education or experience, and are trained for specific assigned duties, (3) raw materials that are traceable, and characterized and qualified for use, and (4) a process (including production, purification, filling, and storage) that is controlled, documented, aseptic, reliable, and consistent, for producing a product that is safe, pure, and potent (see Figs 2.1 and 2.2).

The first component of a successful clinical trial manufacturing program is the area or facility used for the production of clinical trial batches. This facility should have limited access and be segregated from other activities where the danger of cross contamination is removed and the control of materials, product, waste, and personnel flow can be achieved. Preparation of vector batches requires rooms where a controlled environment can be maintained. Here, under class 10,000 and class 100 environments, batches can be produced, purified, and vialed for use in trials. HVAC and other equipment used within the manufacturing facility requires annual calibration, initial installation, and operational qualification, and appropriate preventive maintenance programs. Written procedures for the use, cleaning, and maintenance of the facility, and the equipment must also be written and followed by appropriate personnel.

The second component of a successful clinical manufacturing program is appropriate quality assurance systems for the production of cell or viral banks, raw materials, in-process materials, and final product. Much has been written about the production and testing of cell and viral banks for use in manufacturing. Both FDA guidelines and other reference information are available (http://www.fda.gov.cber/guidelines.htm; http://www.ich.org, http://www.emea.eu.int). The production of AAV vector batches for use in clinical trials should use qualified cell, viral, and/or plasmid banks as appropriate. Documentation on the source materials for these banks is also crucial in assuring the quality and safety of the clinical trial materials.

The materials used in the production process and the vendors producing them must be carefully selected. These raw materials, such as tissue culture media, cell culture plastic ware, water, chemicals, enzymes, or other reagents needed to produce the vector must be evaluated in the process to determine their acceptability for use. Many research-grade reagents are not acceptable for preparation of clinical trial materials; USP or multi-compendial grade chemicals should be used if available. Water for injection, or WFI, is the most appropriate type of water for certain process steps, particularly in the final product preparation. Materials made from or containing ingredients of animal origin should be avoided, however, if use of these types of materials is necessary, certificates of origin for the animal derived components must be obtained from the vendor to ensure acceptability. The requirements for each of the raw materials, components, containers, etc. used in the process must have written specifications, which describe their acceptability criteria. Qualification of raw materials by quality control begins with review of vendor certificates of analysis and matures to actual raw materials testing by Phase 3. A quality assurance administered program of quarantine and release of received raw material lots prior to use in production adds another level of assurance that the final product produced is free of contaminants and safe for use in humans.

Personnel need to have appropriate education and experience, specific responsibilities, job specific training, as well as training in the Current Good Manufacturing Practices. Personnel background information and training should be documented prior to manufacturing. Periodic training on existing documents and documentation of training is an on-going process as procedures are updated and new ones are introduced.

Appropriate documentation of the process methods and the actual reagents, components and supplies used to prepare each clinical batch can be achieved through the use of batch records. These documents capture contemporaneously all activities and materials used in the production of the batch, including any deviations or changes necessary during actual production. Deviations are investigated and explained as to reason and outcome; if needed, corrective actions can be applied and the batch record revised as necessary prior to its next use. The use of batch records also allows for batch identification (lot number assignment) and labeling, so no confusion exists as to the identity of each batch of vector produced. Batch records also incorporate the sampling of the production material as it is processed. In-process and final product sampling is necessary for determining the acceptability of the batch, but also supplies important information on the production process, its reproducibility, and helps identify areas where better control or additional production steps may be required as the product moves through the clinical development process. Testing of the in-process and final product materials from each lot is accomplished through the use of standardized assay methods. Guidelines related to appropriate assays for gene transfer vectors recommend the testing listed in Table 2.3 and outlined in Figs 2.1 and 2.2.

The development of the assay method should concern itself with the assay's intended use. The criticality of the parameter being measured should be the driving force behind decisions surrounding the test method; guidance can be found in the US Pharmacopeia (USP). Questions such as does the assay need to be quantitative or qualitative, what range of quantitation is required, what degree of specificity is needed, etc. should be considered during development. At the same time, appropriate standards or controls should be prepared and evaluated for use during clinical lot testing. Documentation concerning the development of any in-house standard should be as complete as possible. Ideally, data should be collected and compiled to document the acceptability of the assay for its intended use prior to the testing of the first clinical batch. Assay

TABLE 2.3

AAV clinical batch testing

Sterility including bacteriostasis/fungistasis[a,c]	Purity: silver staining/coomassie blue[b,c]
Mycoplasma[a]	Potency: infectious unit titer[b,c]
Endotoxin[c]	Strength: vector genome titer[b,c]
In vitro adventitious viral contaminants[a]	Strength: capsid particle titer[c]
General safety[c]	rcAAV[b]
Appearance[c]	Residual host cell DNA[b]
Identity[c]	Transgene expression[b]
Stability[c]	

[a]Performed on AAV harvest.
[b]Performed on purified bulk.
[c]Performed on vialed final product.

qualification studies and their documentation would proceed to complete assay validation as the product approached Phase 3 clinical trials.

Actual batch testing by quality control is performed under cGMPs and documented using a standardized document such as a test record. This document allows for the capture of all critical information concerning the performance of the test, including the equipment and reagents used to perform the test, the results of the standards and controls, the results from the testing of the clinical batch, the decision concerning the validity of the assay (which is based on the outcome of the standards and controls), and finally the acceptability of the clinical batch results related to the test specifications. Development of such documents makes the capture of all the required information much easier for the analyst performing the test, and ensures that the information can be easily archived.

The development of in-process and final product release specifications is an activity that must rely on experience with the production of the clinical batch. Ideally, trial batches made using the equipment and materials identical to those used for the clinical batches should be tested beforehand by QC. The specification for the vector preparation is dependent on what is regulatorily acceptable (i.e., passing a sterility test) and what the process is capable of producing. For Phase 1 clinical materials, it may be necessary to rely on data generated in the research laboratory using different equipment to establish an expected or needed specification.

The last component of a successful clinical manufacturing program is a quality assurance unit. The QAU must be separate from and independent of personnel responsible for or engaged in the manufacture and testing of the clinical trial material. This unit can provide document control systems for standard operating procedures, batch and test records, raw material and product specifications, etc. GMPs also require that such a unit be responsible for the release of raw materials for manufacturing, and product for clinical use. QA is an independent decision maker concerning the acceptability of product for use in the clinic. This separation adds a high level of assurance that the batch was manufactured by qualified personnel according to documented procedures, in an appropriate environment, using qualified equipment, and was tested appropriately to meet all quality control test specifications. This, of course, is intended to ensure the purity, potency, and safety of the clinical material since all of this information, the process, the testing and specifications, the facility, and the quality systems are all described in the IND, which is submitted to, and reviewed and approved by CBER.

Acknowledgments

Supported by grants from NIH (RR16586, HL51811, HL59412, DK58327, EY13729), the Alpha One Foundation, We acknowledge: Mark Potter, Melissa Breedlove, Glenn Philipsburg, Thomas Andresen, Kristin Good, Irina Korytov, Denna Worley, Tina Philipsberg, Jason Fife, Irina Anikina, Sergei Zolotukhin, Jenna Davis, Irina Zolotukhin, Brian Cleaver, Connie Nicklin, Carolyn Baum, and Nicholas Muzyczka. RS may be entitled to royalties on technology discussed in this article. RS owns equity in a gene therapy company that is commercializing AAV for gene therapy applications.

References

Allen, J. M., Debelak, D. J., Reynolds, T. C. and Miller, A. D. (1997). Identification and elimination of replication-competent adeno-associated virus (AAV) that can arise by nonhomologous recombination during AAV vector production. J. Virol. *71*, 6816–6822.

Arella, M., Garzon, S., Bergeron, J. and Tijssen, P. (1990). Physiochemical properties, production, and purification of parvoviruses. In: Handbook of Parvoviruses (Tijssen, P., ed.), Vol. 1. CRC Press, Boca Raton, pp. 11–30.

Atchison, R. W., Casto, B. C. and Hammon, W. M. (1966). Electron microscopy of adenovirus-associated virus (AAV) in cell cultures. Virology *29*, 353–357.

Carter, B. J. (1990). The growth cycle of adeno-associated virus. In: Handbook of Parvoviruses (Tijssen, P., ed.), Vol. 1. CRC Press, Boca Raton, pp. 155–168.

CBER (1998). Guidance for human somatic cell therapy and gene therapy. Hum. Gene Ther. *9*, 1513–1524.

Chadeuf, G. et al. (2000). Efficient recombinant adeno-associated virus production by a stable rep-cap HeLa cell line correlates with adenovirus-induced amplification of the integrated rep-cap genome. J. Gene Med. *2*, 260–268.

Chang, L. S. and Shenk, T. (1990). The adenovirus DNA-binding protein stimulates the rate of transcription directed by adenovirus and adeno-associated virus promoters. J. Virol. *64*, 2103–2109.

Chao, H. et al. (2000). Several log increase in therapeutic transgene delivery by distinct adeno-associated viral serotype vectors. Mol. Ther. *2*, 619–623.

Chejanovsky, N. and Carter, B. J. (1989). Mutagenesis of an AUG codon in the adeno-associated virus rep gene: Effects on viral DNA replication. Virology *173*, 120–128.

Chiorini, J. A., Kim, F., Yang, L. and Kotin, R. M. (1999). Cloning and characterization of adeno-associated virus type 5. J. Virol. *73*, 1309–1319.

Chiorini, J. A., Yang, L., Liu, Y., Safer, B. and Kotin, R. M. (1997). Cloning of adeno-associated virus type 4 (AAV4) and generation of recombinant AAV4 particles. J. Virol. *71*, 6823–6833.

Chiorini, J. A. et al. (1998). Inhibition of PrKX, a novel protein kinase, and the cyclic AMP-dependent protein kinase PKA by the regulatory proteins of adeno-associated virus type 2. Mol. Cell. Biol. *18*, 5921–5929.

Clark, K. R., Liu, X., McGrath, J. P. and Johnson, P. R. (1999). Highly purified recombinant adeno-associated virus vectors are biologically active and free of detectable helper and wild-type viruses [In Process Citation]. Hum. Gene Ther. *10*, 1031–1039.

Clark, K. R., Voulgaropoulou, F., Fraley, D. M. and Johnson, P. R. (1995). Cell lines for the production of recombinant adeno-associated virus. Hum. Gene Ther. *6*, 1329–1341.

Clark, K. R., Voulgaropoulou, F. and Johnson, P. R. (1996). A stable cell line carrying adenovirus-inducible rep and cap genes allows for infectivity titration of adeno-associated virus vectors. Gene Ther. *3*, 1124–1132.

Conway, J. E., Zolotukhin, S., Muzyczka, N., Hayward, G. S. and Byrne, B. J. (1997). Recombinant adeno-associated virus type 2 replication and packaging is entirely supported by a herpes simplex virus type 1 amplicon expressing Rep and Cap. J. Virol. *71*, 8780–8789.

Conway, J. E. et al. (1999). High-titer recombinant adeno-associated virus production utilizing a recombinant herpes simplex virus type I vector expressing AAV-2 Rep and Cap. Gene Ther. *6*, 986–993.

Davidson, B. L. et al. (2000). Recombinant adeno-associated virus type 2, 4, and 5 vectors: Transduction of variant cell types and regions in the mammalian central nervous system. Proc. Natl. Acad. Sci. USA *97*, 3428–3432.

Di Pasquale, G. and Stacey, S. N. (1998). Adeno-associated virus Rep78 protein interacts with protein kinase A and its homolog PRKX and inhibits CREB-dependent transcriptional activation. J. Virol. *72*, 7916–7925.

Di Pasquale, G., Davidson, B. L., Stein, C. S., Martins, I., Scudiero, D., Monks, A. and Chiorini, J. A. (2003). Identification of PDGFR as a receptor for AAV-5 transduction. Nat. Med. *9*, 1306–1312.

Ferrari, F. K., Samulski, T., Shenk, T. and Samulski, R. J. (1996). Second-strand synthesis is a rate-limiting step for efficient transduction by recombinant adeno-associated virus vectors. J. Virol. *70*, 3227–3234.

Ferrari, F. K., Xiao, X., McCarty, D. and Samulski, R. J. (1997). New developments in the generation of Ad-free, high-titer rAAV gene therapy vectors. Nat. Med. *3*, 1295–1297.

Flotte, T. et al. (1996). A phase I study of an adeno-associated virus-CFTR gene vector in adult CF patients with mild lung disease. Hum. Gene Ther. *7*, 1145–1159.

Gao, G. P. et al. (1998). High-titer adeno-associated viral vectors from a Rep/Cap cell line and hybrid shuttle virus [In Process Citation]. Hum. Gene Ther. *9*, 2353–2362.

Gao, G. P. et al. (2002). Novel adeno-associated viruses from rhesus monkeys as vectors for human gene therapy. Proc. Natl. Acad. Sci. USA *99*, 11854–11859.

Gao, G. et al. (2003). Adeno-associated viruses undergo substantial evolution in primates during natural infections. Proc. Natl. Acad. Sci. USA *100*, 6081–6086.

Grimm, D. (2002). Production methods for gene transfer vectors based on adeno-associated virus serotypes. Methods *28*, 146–157.

Grimm, D., Kern, A., Rittner, K. and Kleinschmidt, J. A. (1998). Novel tools for production and purification of recombinant adenoassociated virus vectors [In Process Citation]. Hum. Gene Ther. *9*, 2745–2760.

Grimm, D. et al. (1999). Titration of AAV-2 particles via a novel capsid ELISA: Packaging of genomes can limit production of recombinant AAV-2. Gene Ther. *6*, 1322–1330.

Halbert, C. L., Allen, J. M. and Miller, A. D. (2001). Adeno-associated virus type 6 (AAV6) vectors mediate efficient transduction of airway epithelial cells in mouse lungs compared to that of AAV2 vectors. J. Virol. *75*, 6615–6624.

Halbert, C. L., Rutledge, E. A., Allen, J. M., Russell, D. W. and Miller, A. D. (2000). Repeat transduction in the mouse lung by using adeno-associated virus vectors with different serotypes. J. Virol. *74*, 1524–1532.

Hermonat, P. L., Labow, M. A., Wright, R., Berns, K. I. and Muzyczka, N. (1984). Genetics of adeno-associated virus: Isolation and preliminary characterization of adeno-associated virus type 2 mutants. J. Virol. *51*, 329–339.

Hermonat, P. L. and Muzyczka, N. (1984). Use of adeno-associated virus as a mammalian DNA cloning vector: Transduction of neomycin resistance into mammalian tissue culture cells. Proc. Natl. Acad. Sci. USA *81*, 6466–6470.

Hildinger, M. et al. (2001). Hybrid vectors based on adeno-associated virus serotypes 2 and 5 for muscle-directed gene transfer. J. Virol. *75*, 6199–6203.

Horer, M. et al. (1995). Mutational analysis of adeno-associated virus Rep protein-mediated inhibition of heterologous and homologous promoters. J. Virol. *69*, 5485–5496.

Huser, D., Weger, S. and Heilbronn, R. (2002). Kinetics and frequency of adeno-associated virus site-specific integration into human chromosome 19 monitored by quantitative real-time PCR. J. Virol. *76*, 7554–7559.

Im, D. S. and Muzyczka, N. (1990). The AAV origin binding protein Rep68 is an ATP-dependent site-specific endonuclease with DNA helicase activity. Cell *61*, 447–457.

Janik, J. E., Huston, M. M., Cho, K. and Rose, J. A. (1989). Efficient synthesis of adeno-associated virus structural proteins requires both adenovirus DNA binding protein and VA I RNA. Virology *168*, 320–329.

Janson, C. et al. (2002). Clinical protocol. Gene therapy of Canavan disease: AAV-2 vector for neurosurgical delivery of aspartoacylase gene (ASPA) to the human brain. Hum. Gene Ther. *13*, 1391–1412.

Jay, F. T., Laughlin, C. A. and Carter, B. J. (1981). Eukaryotic translational control: Adeno-associated virus protein synthesis is affected by a mutation in the adenovirus DNA-binding protein. Proc. Natl. Acad. Sci. USA *78*, 2927–2931.

Kaludov, N., Brown, K. E., Walters, R. W., Zabner, J. and Chiorini, J. A. (2001). Adeno-associated virus serotype 4 (AAV4) and AAV5 both require sialic acid binding for hemagglutination and efficient transduction but differ in sialic acid linkage specificity. J. Virol. *75*, 6884–6893.

Kaludov, N. et al. (2003). Production, purification and preliminary X-ray crystallographic studies of adeno-associated virus serotype 4. Virology *306*, 1–6.

Kay, M. A. et al. (2000). Evidence for gene transfer and expression of factor IX in haemophilia B patients treated with an AAV vector. Nat. Genet. *24*, 257–261.

King, J. A., Dubielzig, R., Grimm, D. and Kleinschmidt, J. A. (2001). DNA helicase-mediated packaging of adeno-associated virus type 2 genomes into preformed capsids. EMBO J. *20*, 3282–3291.

Klein, R. L. et al. (1998). Neuron-specific transduction in the rat septohippocampal or nigrostriatal pathway by recombinant adeno-associated virus vectors. Exp. Neurol. *150*, 183–194.
Kotin, R. M. et al. (1990). Site-specific integration by adeno-associated virus. Proc. Natl. Acad. Sci. USA *87*, 2211–2215.
Krieg, A. M., Yi, A. K., Schorr, J. and Davis, H. L. (1998). The role of CpG dinucleotides in DNA vaccines. Trends Microbiol. *6*, 23–27.
Labow, M. A., Hermonat, P. L. and Berns, K. I. (1986). Positive and negative autoregulation of the adeno-associated virus type 2 genome. J. Virol. *60*, 251–258.
Li, J., Samulski, R. J. and Xiao, X. (1997). Role for highly regulated rep gene expression in adeno-associated virus vector production. J. Virol. *71*, 5236–5243.
Liu, X. L., Clark, K. R. and Johnson, P. R. (1999). Production of recombinant adeno-associated virus vectors using a packaging cell line and a hybrid recombinant adenovirus. Gene Ther. *6*, 293–299.
Mah, C. et al. (1998). Adeno-associated virus type 2-mediated gene transfer: Role of epidermal growth factor receptor protein tyrosine kinase in transgene expression. J. Virol. *72*, 9835–9843.
Manno, C. S. et al. (2003). AAV-mediated factor IX gene transfer to skeletal muscle in patients with severe hemophilia B. Blood *101*, 2963–2972.
Mathews, L. C., Gray, J. T., Gallagher, M. R. and Snyder, R. O. (2002). Recombinant adeno-associated viral vector production using stable packaging and producer cell lines. Methods Enzymol. *346*, 393–413.
Matsushita, T. et al. (1998). Adeno-associated virus vectors can be efficiently produced without helper virus. Gene Ther. *5*, 938–945.
McCarty, D. M., Christensen, M. and Muzyczka, N. (1991). Sequences required for coordinate induction of adeno-associated virus p19 and p40 promoters by Rep protein. J. Virol. *65*, 2936–2945.
McLaughlin, S. K., Collis, P., Hermonat, P. L. and Muzyczka, N. (1988). Adeno-associated virus general transduction vectors: Analysis of proviral structures. J. Virol. *62*, 1963–1973.
Nakai, H. et al. (2001). Extrachromosomal recombinant adeno-associated virus vector genomes are primarily responsible for stable liver transduction in vivo. J. Virol. *75*, 6969–6976.
Owens, R. A., Weitzman, M. D., Kyostio, S. R. and Carter, B. J. (1993). Identification of a DNA-binding domain in the amino terminus of adeno-associated virus Rep proteins. J. Virol. *67*, 997–1005.
Pereira, D. J. and Muzyczka, N. (1997). The cellular transcription factor SP1 and an unknown cellular protein are required to mediate Rep protein activation of the adeno-associated virus p19 promoter. J. Virol. *71*, 1747–1756.

Pilder, S., Moore, M., Logan, J. and Shenk, T. (1986). The adenovirus E1B-55K transforming polypeptide modulates transport or cytoplasmic stabilization of viral and host cell mRNAs. Mol. Cell. Biol. *6*, 470–476.

Qiao, C., Li, J., Skold, A., Zhang, X. and Xiao, X. (2002a). Feasibility of generating adeno-associated virus packaging cell lines containing inducible adenovirus helper genes. J. Virol. *76*, 1904–1913.

Qiao, C., Wang, B., Zhu, X., Li, J. and Xiao, X. (2002b). A novel gene expression control system and its use in stable, high-titer 293 cell-based adeno-associated virus packaging cell lines. J. Virol. *76*, 13015–13027.

Qing, K. et al. (1999). Human fibroblast growth factor receptor 1 is a co-receptor for infection by adeno-associated virus 2 [In Process Citation]. Nat. Med. *5*, 71–77.

Rabinowitz, J. E. et al. (2002). Cross-packaging of a single adeno-associated virus (AAV) type 2 vector genome into multiple AAV serotypes enables transduction with broad specificity. J. Virol. *76*, 791–801.

Richardson, W. D. and Westphal, H. (1981). A cascade of adenovirus early functions is required for expression of adeno-associated virus. Cell *27*, 133–141.

Rutledge, E. A., Halbert, C. L. and Russell, D. W. (1998). Infectious clones and vectors derived from adeno-associated virus (AAV) serotypes other than AAV type 2. J. Virol. *72*, 309–319.

Salvetti, A. et al. (1998). Factors influencing recombinant adeno-associated virus production. Hum. Gene Ther. *9*, 695–706.

Samulski, R. J., Berns, K. I., Tan, M. and Muzyczka, N. (1982). Cloning of adeno-associated virus into pBR322: Rescue of intact virus from the recombinant plasmid in human cells. Proc. Natl. Acad. Sci. USA *79*, 2077–2081.

Samulski, R. J., Chang, L. S. and Shenk, T. (1989). Helper-free stocks of recombinant adeno-associated viruses: Normal integration does not require viral gene expression. J. Virol. *63*, 3822–3828.

Samulski, R. J. et al. (1991). Targeted integration of adeno-associated virus (AAV) into human chromosome 19 [published erratum appears in EMBO J. 1992 Mar;11(3):1228]. EMBO J. *10*, 3941–3950.

Schnepp, B. C., Clark, K. R., Klemanski, D. L., Pacak, C. A. and Johnson, P. R. (2003). Genetic fate of recombinant adeno-associated virus vector genomes in muscle. J. Virol. *77*, 3495–3504.

Shi, Y., Seto, E., Chang, L. S. and Shenk, T. (1991). Transcriptional repression by YY1, a human GLI-Kruppel-related protein, and relief of repression by adenovirus E1A protein. Cell *67*, 377–388.

Smith, R. H. and Kotin, R. M. (1998). The Rep52 gene product of adeno-associated virus is a DNA helicase with $3'$-to-$5'$ polarity. J. Virol. *72*, 4874–4881.

Smith, R. H., Spano, A. J. and Kotin, R. M. (1997). The Rep78 gene product of adeno-associated virus (AAV) self-associates to form a hexameric complex in the presence of AAV ori sequences. J. Virol. *71*, 4461–4471.

Snyder, R. O., Xiao, X. and Samulski, R. J. (1996). Production of recombinant adeno-associated viral vectors. In: Current Protocols in Human Genetics (Dracopoli, N., ed.). John Wiley and Sons, New York, pp. 12.1.1–24.

Srivastava, A., Lusby, E. W. and Berns, K. I. (1983). Nucleotide sequence and organization of the adeno-associated virus 2 genome. J. Virol. *45*, 555–564.

Stedman, H., Wilson, J. M., Finke, R., Kleckner, A. L. and Mendell, J. (2000). Phase I clinical trial utilizing gene therapy for limb girdle muscular dystrophy: Alpha-, beta-, gamma-, or delta-sarcoglycan gene delivered with intramuscular instillations of adeno-associated vectors. Hum. Gene Ther. *11*, 777–790.

Summerford, C., Bartlett, J. S. and Samulski, R. J. (1999). AlphaVbeta5 integrin: A co-receptor for adeno-associated virus type 2 infection [In Process Citation]. Nat. Med. *5*, 78–82.

Summerford, C. and Samulski, R. J. (1998). Membrane-associated heparan sulfate proteoglycan is a receptor for adeno-associated virus type 2 virions. J. Virol. *72*, 1438–1445.

Sun, B. et al. (2003). Packaging of an AAV vector encoding human acid alpha-glucosidase for gene therapy in glycogen storage disease type II with a modified hybrid adenovirus-AAV vector. Mol. Ther. *7*, 467–477.

Tratschin, J. D., Miller, I. L. and Carter, B. J. (1984). Genetic analysis of adeno-associated virus: Properties of deletion mutants constructed in vitro and evidence for an adeno-associated virus replication function. J. Virol. *51*, 611–619.

Tratschin, J. D., West, M. H., Sandbank, T. and Carter, B. J. (1984). A human parvovirus, adeno-associated virus, as a eucaryotic vector: Transient expression and encapsidation of the procaryotic gene for chloramphenicol acetyltransferase. Mol. Cell. Biol. *4*, 2072–2081.

Urabe, M., Ding, C. and Kotin, R. M. (2002). Insect cells as a factory to produce adeno-associated virus type 2 vectors. Hum. Gene Ther. *13*, 1935–1943.

Urcelay, E., Ward, P., Wiener, S. M., Safer, B. and Kotin, R. M. (1995). Asymmetric replication in vitro from a human sequence element is dependent on adeno-associated virus Rep protein. J. Virol. *69*, 2038–2046.

Wagner, J. A. et al. (1998). A phase I/II study of tgAAV-CF for the treatment of chronic sinusitis in patients with cystic fibrosis. Hum. Gene Ther. *9*, 889–909.

Wagner, J. A. et al. (2002). A phase II, double-blind, randomized, placebo-controlled clinical trial of tgAAVCF using maxillary sinus delivery in patients with cystic fibrosis with antrostomies. Hum. Gene Ther. *13*, 1349–1359.

Walters, R. W. et al. (2001). Binding of adeno-associated virus type 5 to 2,3-linked sialic acid is required for gene transfer. J. Biol. Chem. *276*, 20610–20616.

Wang, X. S. et al. (1998). Characterization of wild-type adeno-associated virus type 2-like particles generated during recombinant viral vector production and strategies for their elimination. J. Virol. *72*, 5472–5480.

Weger, S., Wendland, M., Kleinschmidt, J. A. and Heilbronn, R. (1999). The adeno-associated virus type 2 regulatory proteins rep78 and rep68 interact with the transcriptional coactivator PC4. J. Virol. *73*, 260–269.

Weindler, F. W. and Heilbronn, R. (1991). A subset of herpes simplex virus replication genes provides helper functions for productive adeno-associated virus replication. J. Virol. *65*, 2476–2483.

Weitzman, M. D., Kyostio, S. R., Kotin, R. M. and Owens, R. A. (1994). Adeno-associated virus (AAV) Rep proteins mediate complex formation between AAV DNA and its integration site in human DNA. Proc. Natl. Acad. Sci. USA *91*, 5808–5812.

Wu, P. et al. (2000). Mutational analysis of the adeno-associated virus type 2 (AAV2) capsid gene and construction of AAV2 vectors with altered tropism. J. Virol. *74*, 8635–8647.

Xiao, W. et al. (1999). Gene therapy vectors based on adeno-associated virus type 1. J. Virol. *73*, 3994–4003.

Xiao, X., Li, J. and Samulski, R. J. (1998). Production of high-titer recombinant adeno-associated virus vectors in the absence of helper adenovirus. J. Virol. *72*, 2224–2232.

Xie, Q. et al. (2002). The atomic structure of adeno-associated virus (AAV-2), a vector for human gene therapy. Proc. Natl. Acad. Sci. USA *99*, 10405–10410.

Zolotukhin, S., Potter, M., Hauswirth, W. W., Guy, J. and Muzyczka, N. (1996). A "humanized" green fluorescent protein cDNA adapted for high-level expression in mammalian cells. J. Virol. *70*, 4646–4654.

Zolotukhin, S. et al. (1999). Recombinant adeno-associated virus purification using novel methods improves infectious titer and yield. Gene Ther. *6*, 973–985.

Zolotukhin, S. et al. (2002). Production and purification of serotype 1, 2, and 5 recombinant adeno-associated viral vectors. Methods *28*, 158–167.

CHAPTER 3

Gene therapy for hemophilia

Cathryn Mah

Department of Pediatrics, Division of Cell and Molecular Therapies, University of Florida, Box 100296, Gainesville, FL 32610-0296, USA

The initial gene transfer studies for the treatment of bleeding disorders were performed using retrovirus-based vectors in the late 1980s. Since the first approved gene therapy trial in the US in 1990, the field of gene therapy research has exploded, resulting in an ever-growing multitude of vector systems. Not only have there been significant advances in the understanding of vector biology, but the field of gene therapy has also pushed toward a greater understanding of the underlying mechanisms and etiology of a broad range of human disease.

Within the past 5 years, five human gene therapy clinical trials for bleeding disorders have been initiated (Roth et al., 2001; Kelley et al., 2002; High, 2003; Manno et al., 2003; Powell et al., in press). The first clinical trial for bleeding disorders commenced in 1998 and was for the treatment of hemophilia A (Roth et al., 2001). To date, most gene therapy studies have focused on the treatment of hemophilia A or B. According to the World Federation of Hemophilia approximately 1 in 10,000 males born worldwide have hemophilia. Hemophilia A is an X-linked disorder that arises from a lack of functional clotting factor VIII and hemophilia B is caused by a deficiency of functional clotting factor IX. A schematic shown in Fig. 3.1 depicts the roles that factor VIII and IX play in

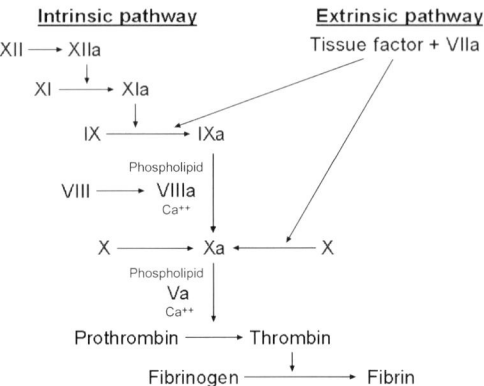

Fig. 3.1. Schematic of the intrinsic and extrinsic pathways of the coagulation cascade.

the blood coagulation cascade. Hemophilia A constitutes 80% of all hemophilia cases (Hoyer, 1994; Connelly and Kaleko, 1998; Kelley et al., 2002; High, 2003).

Currently, hemophilia is treated with protein replacement therapy. For most patients, treatment occurs in response to bleeding episodes rather than as a prophylactic measure. Protein replacement therapy is limited by the high cost of production of purified recombinant protein, the short half-life of clotting factors, the requirement for repeated intravenous administration, and the development of host antibodies to the protein. The use of somatic cell gene therapy may circumvent some of these problems in that constant production of the deficient clotting factor in vivo could be achieved. In the case of hemophilia A, as little as 2% of the normal levels of factor VIII can result in therapeutic benefits (Connelly et al., 1999; Herzog et al., 1999; Vandendriessche et al., 1999; Sarkar et al., 2000; Mah et al., 2003; Mingozzi et al., 2003). Gene therapy strategies may be ideal for the treatment of hemophilia and other bleeding disorders, in that disease results from mutations of single genes that have been well characterized and a wealth of information exists concerning the effects of therapeutic recombinant proteins on affected individuals. Furthermore, both genetically engineered and naturally occurring animal models of hemophilia are available (Evans et al., 1989; Bi et al., 1995, 1996; Lin et al., 1997; Hough et al., 2002; Lozier et al., 2002). As a result, five different vector systems are currently being investigated for their potential use as gene therapy vectors for the treatment of hemophilia. These include non-viral DNA, adenovirus, retrovirus, lentivirus, and adeno-associated virus (AAV) vectors.

3.1. Non-viral DNA vectors

Advantages of non-viral DNA vectors include unrestricted gene length and little vector-associated pathology. Pre-clinical studies using non-viral vectors have demonstrated sustained expression of both factor VIII and factor IX in murine models (Yant, 2000; Miao et al., 2001; Lin et al., 2002). Historically, one disadvantage of non-viral DNA vector systems has been the lack of persistence of the therapeutic transgene in vivo. Improvements such as the development of a Sleeping Beauty transposon-based system or the incorporation of viral sequences to promote integration into the host chromosome have been developed to address this issue (Yant, 2000). For all non-viral vector systems, the method of delivery is important because it is unlike viral vectors in which the viral particle itself takes part in the transduction of cells. In murine pre-clinical studies for the treatment of hemophilia B, a hydrodynamic approach for the delivery of plasmid DNA has been successfully used. In these studies, approximately 20 μg plasmid DNA in 2 ml saline is injected directly into the tail vein of adult mice over 5–8 s, resulting in significant transduction of the liver (Yant, 2000; Miao et al., 2001). This hydrodynamic delivery approach is a simple and relatively noninvasive method to target liver transduction in murine models, but may not be as useful a method for human application. Other, more clinically applicable approaches involve ex vivo transfection followed by implantation of transduced cells in vivo. An ex vivo approach is attractive because only the desired target cells are isolated and transduced, thereby reducing the likelihood of potential immune response to the vector or inadvertent transduction of non-target cells. The first human gene therapy clinical trial for bleeding disorders, sponsored by Transkaryotic Therapies, Inc. (Cambridge, MA), uses an ex vivo non-viral DNA vector strategy for the delivery of human factor VIII for the treatment of hemophilia A (Roth et al., 2001). In this phase I safety study, subject fibroblasts are isolated, transfected with the factor VIII gene followed by intraperitoneal implantation. The greatest improvement seen in any of the current phase I gene therapy trials for bleeding disorders was noted in this trial with 4% of normal circulating factor VIII, albeit transient, in one subject. To date, only very modest improvements have been noted in non-viral DNA clinical trial subjects, all of which have been short-term in duration. It is important to note, though, that all trials thus far have been phase I trials aimed at assessing safety profiles of the vectors, rather than aiming for disease correction.

3.2. Adenovirus vectors

There exists a substantial body of work investigating the use of adenovirus-based vectors for the treatment of hemophilia A. Several groups have shown that adenoviral vectors yield very high levels of factor VIII expression and although these vectors do not integrate into the host chromosome, some groups have shown sustained expression in both hemophilia A mouse and dog model studies (Connelly et al., 1999; Sarkar et al., 2000; Andrews et al., 2001; Bristol et al., 2001; Gallo-Penn et al., 2001). In recent years, the immunology surrounding viral-based gene therapy vector systems, and in particular, adenovirus-based systems, has been under intense scrutiny. Adenoviral vectors historically have resulted in high transduction efficiency and robust transgene expression but many instances of immune response to the vector and/or transgene have led to more improved versions of the vector system. Currently gutless adenoviral vectors devoid of all viral genes are being examined in pre-clinical studies for the treatment of hemophilia (Balague et al., 2000; Ehrhardt and Kay, 2002; Reddy et al., 2002; Chuah et al., 2003). When administered intravenously, adenoviral vectors have an organ-tropism for the liver. The liver is an ideal target tissue for gene therapy for hemophilia, as both clotting factors VIII and IX are made in the liver in vivo. Furthermore, the liver is well suited for expression of secreted proteins, and the high vascularization of the tissue provides an ideal environment for the secreted protein to easily enter systemic circulation. Several recent studies using gutless adenoviral vectors carrying the human factor VIII gene under the control of liver-specific promoters showed initial normal physiologic, even supraphysiologic levels of circulating factor VIII, when administered to hemophilia A mice. Chuah et al. demonstrated 75,000 mU/ml factor VIII within the first 5–8 days post-injection (Chuah et al., 2003). Similar studies by Reddy et al. and Balague et al. showed approximately 15,000 mU/ml 2 weeks post-administration and up to 1000 ng/ml circulating factor VIII protein, respectively (Balague et al., 2000; Reddy et al., 2002). In the studies by Chuah et al., factor VIII levels in all treated animals fell to baseline by 3–4 weeks and in the study by Balague et al., only 3 of 16 treated mice had sustained expression beyond 100 days, of which one animal had detectable factor levels through more than 1 year. The drop in factor VIII levels in all mice correlated with the formation of inhibitory antibodies to the human factor VIII protein. Conversely, in the study by Reddy et al., detectable factor VIII levels were seen in the treated animals out to 10 months and there was no evidence of factor VIII-specific antibodies. The difference in results may be attributable to several factors as the vector constructs in all three studies differed with regard to the exact promoter used and the nature of the transgene. Furthermore, it has been suggested that the hemophilia A mouse colonies may have undergone some genetic drift, causing their response to each vector to vary. Using the vector described in the work by Balague et al., GenStar Therapeutics (San Diego, CA) initiated a phase I clinical trial in June 2001 (Balague et al., 2000).

In this study, the first subject had reported levels of 3% normal factor VIII activity. However, elevated liver enzymes in this subject resulted in temporary suspension of the trial, which was then allowed to resume at lower vector dosages. This trial was prematurely terminated in early 2003 due to a lack of recruited subjects (Kelley et al., 2002; High, 2003).

3.3. Retrovirus vectors

Retroviral vectors were the first viral-based vectors studied for the treatment of bleeding disorders. Retroviral vectors have the capacity for a large transgene cassette and have the ability to integrate into host chromosome of actively dividing cells. The safety of a replication-deficient Moloney murine leukemia-based vector expressing a B domain-deleted (BDD) version of the human factor VIII gene for the treatment of hemophilia A is being currently assessed in a phase I clinical trial sponsored by Chiron Corporation (Emeryville, CA) (Kelley et al., 2002; High, 2003; Powell et al., in press). Recent results from this trial showed that 8 of 12 treated subjects had greater than 1% activity on two or more occasions during the 53-week period of the study. Interestingly, there was no correlation of vector dose to factor VIII levels or the time at which the expression was first seen (Powell et al., in press). Pre-clinical studies examining the utility of retroviral vectors for the treatment of hemophilia B are also currently underway (Vandendriessche et al., 1999; Xu et al., 2003). Work done by Xu et al. also used a replication-deficient Moloney murine leukemia virus-based vector to deliver the human factor IX gene under the control of the liver-specific alpha-1-antitrypsin promoter. In these studies the vector was administered intravenously to newborn hemophilia B mice or puppies, resulting in an average of 162% or 2.5–6.1% normal levels of factor IX, respectively, of which some animals expressed therapeutic levels out to 11 months (Xu et al., 2003). Interesting advantages to delivery of vector to neonates is that this method can result in a higher effective dose per kilogram body weight and, in the case of retroviral vectors, the liver, is more actively dividing than in adults, increasing the likelihood of vector integration and persistence. Several groups have also investigated the use of retroviral vectors in ex vivo approaches (Dwarki et al., 1995; Hao et al., 1995; Chuah et al., 2000; Krebsbach et al., 2003; Van Damme et al., 2003). Advantages of ex vivo gene therapy strategies include the ability to manipulate in vitro conditions to maximize transduction efficiencies and to control which cell population is transduced, thereby avoiding potential unwanted promiscuous transduction events. Most recently, studies by Krebsbach et al. and Van Damme et al. used retroviral vectors to deliver the factor IX or factor VIII genes ex vivo to bone marrow stem cells or bone marrow mesenchymal cells, respectively, followed by implantation into immunocompetent mice. Krebsbach et al. observed factor IX protein expression out to 16 weeks, with a maximum of 25 ng/ml 1 week post-implantation (Krebsbach et al., 2003; Van Damme et al., 2003). To address the potential for graft rejection in ex vivo strategies, a modified strategy involving the implantation of transduced cells encapsulated in material permeable to the expressed clotting factor but impermeable to host immune mediators, has also been described (Hortelano et al., 2001; Garcia-Martin et al., 2002). Currently, most retroviral vector-based ex vivo studies for bleeding disorders have been performed with heterologous donor cells, which may have been instrumental in potentiating immune responses. Use of the subject's own cells, similar to the strategy employed in the non-viral DNA clinical trial by Transkaryotic Therapies, Inc., may reduce the potential for such an immune response.

A newer generation of retroviral vectors, based on the lentivirus, hold the same advantages of retroviral vectors with the added ability to transduce non-dividing cells efficiently. Pre-clinical studies with lentiviral vectors for the treatment of hemophilia have shown great promise. Using a feline immunodeficiency virus (FIV)-based vector, Stein et al. showed that 9 of 12 treated hemophilia A mice could express greater than 5 ng/ml factor VIII, six of which expressed factor VIII throughout the duration of the study, which ranged between 101 and 155 days (Stein et al., 2001). Similarly, Park et al. demonstrated that 15% normal human factor IX levels could be achieved in adult C57BL/6 mice (Park et al., 2000). Finally, Kootstra et al. were able to detect low levels of factor VIII in hemophilia A mice out to 3 months using an HIV-1 based vector (Kootstra et al., 2003). As lentiviral vectors are still in the developmental stages, future work regarding lentiviral vector safety, mechanisms of transgene expression, and vector production should lead to the development of more effective gene therapy vectors for bleeding disorders.

3.4. Adeno-associated virus vectors

In the recent years, much focus has been placed on the development of adeno-associated virus-based vectors for the treatment of bleeding disorders. AAV is not associated with any known pathology and, like retrovirus vectors, can integrate into the host chromosome. AAV vectors are currently being assessed in two phase I clinical trials sponsored by Avigen, Inc. (Alameda, CA) for the treatment of hemophilia B (Kay et al., 2000; Kelley et al., 2002; High, 2003; Manno et al., 2003). The first trial focuses on the intramuscular administration of an AAV2 vector carrying the human factor IX gene. In the second trial, AAV2-factor IX vector is administered via intrahepatic artery injection. Pre-clinical studies have shown dramatic sustained levels of factor IX expression in both murine and dog models of hemophilia B using AAV2 vectors administered either intramuscularly or intrahepatically. Recently, reports from the first clinical trial provide evidence of viral transduction, although most subjects had less than 1% circulating factor IX (Kay et al., 2000; Manno et al., 2003). From a safety standpoint, the results of this trial are promising in that no significant toxicities or adverse events were observed. Manno et al. suggest that higher vector doses or methods in which to increase transduction efficiency may be required to achieve therapeutic levels of AAV-mediated muscle-derived factor IX expression (Manno et al., 2003). The promising data from the intramuscular trial served to initiate the second currently ongoing AAV2/hemophilia B clinical trial, in which vector is directed to the liver. The liver is an ideal target organ for factor IX gene therapy in particular because it serves normally as the depot for factor IX production in vivo and is capable of correctly processing and secreting the expressed protein into systemic circulation.

AAV-mediated gene therapy strategies for the treatment of hemophilia A are more complex than for factor IX therapy. One disadvantage of AAV as a gene therapy vector is the limited carrying capacity of approximately 4.7 kb, which necessitates manipulation or truncation of larger genes. The full-length human factor VIII cDNA is approximately 7.2 kb and the BDD version is approximately 4.5 kb. Several novel approaches have been taken for AAV-factor VIII gene therapy. In one approach, very small promoter elements are used to drive BDD factor VIII expression (Gnatenko et al., 1999; Chao et al., 2000a). Chao et al. were able to demonstrate therapeutic levels (2–3% of normal) human factor VIII expression in NOD/SCID and immunocompetent C57BL/6 mice for greater than 11 months, using an intrahepatically administered AAV2-BDD factor VIII vector driven by a hybrid hepatitis B virus enhancer I-herpes simplex thymidine kinase promoter element of approximately 120 bp (Chao et al., 2000a). Recent work by Scallan et al. showed averages of 2.6 and 2.8% normal levels of canine factor VIII over 14 and 7.5 months, respectively, in two hemophilia A dogs administered AAV2 vector carrying a truncated canine factor VIII cDNA driven by a 202 bp region of the transthyretin promoter (Scallan et al., 2003). Limitations on the size of required cis-elements, such as promoter elements, and the concern for developing safe vectors with narrow tissue specificity led to alternative dual-vector approaches for AAV-mediated factor VIII gene therapy. In this method, the heavy and light chain subunit genes of the factor VIII protein are delivered separately in two separate AAV vectors, with the intent of subunit self-assembly in vivo to generate a functional protein. Using this approach, Burton et al. showed approximately 1.2–2.2 IU human factor VIII in C57BL/6 mice coadministered AAV2 vectors carrying the human elongation factor 1α promoter-driven human factor VIII heavy chain and light chain genes, respectively, 8 weeks post-portal vein injection. Similar levels of expression were noted at 5 months post-injection (Burton et al., 1999). Studies by Mah et al. also used the dual-vector approach for the delivery of the murine factor VIII heavy and light chain genes to hemophilia A mice. Intravenous administration to neonates resulted in greater than 6% normal levels of functional factor VIII which was sustained for greater than 5 months. Mah et al. also noted that in order for a dual-vector system to be successful, coinfection of a single cell with both the factor VIII heavy chain and light chain vectors was required, as the heterodimerization event to generate a functional protein did not occur extracellularly. The requirement for coinfection of cells for the generation of functional factor VIII protein may necessitate the use of specialized vectors that specifically target certain cells types or focused methods of vector delivery in order to maximize the efficiency of dual transduction (Mah et al., 2003).

A modified dual-vector strategy takes advantage of the phenomenon of head-to-tail concatamerization of rAAV vector genomes post-entry into the host cell. In this system, two vectors are required: One to carry the promoter element, $5'$-end of the transgene, and a splice donor element, and the other to carry a splice acceptor element and the $3'$-end of the transgene. Following infection of a single cell, the vector genomes concatamerize. Genomes that have concatamerized in a head-to-tail orientation can then be recognized as a pseudogene that can then be spliced and processed to generate a full-length cDNA of the therapeutic gene product. In this system, only those vector genomes that have appropriately concatamerized, that is, $5'$ vector to $3'$ vector in a head-to-tail orientation, can generate protein, eliminating the generation of nonfunctional protein products resulting from the expression of one subunit in the absence of the other necessary subunit and adding another level of safety (Sun et al., 2000; Chao et al., 2002; Duan et al., 2003; Mah et al., 2003;). Using this system, Chao et al. were able to achieve an average of 2% normal human factor VIII levels in NOD/SCID mice that persisted 4 months (Chao et al., 2002). Like the dual-vector system in which the factor VIII subunits are expressed separately, the concatamerization of vectors to generate a therapeutic pseudo-gene also requires coinfection of the same cell with both the vectors. Furthermore, of the four possible concatamer configurations, only those vectors that have concatamerized in a head-to-tail orientation can be processed to generate a functional gene. Schematics of the dual-vector strategies are depicted in Fig. 3.2.

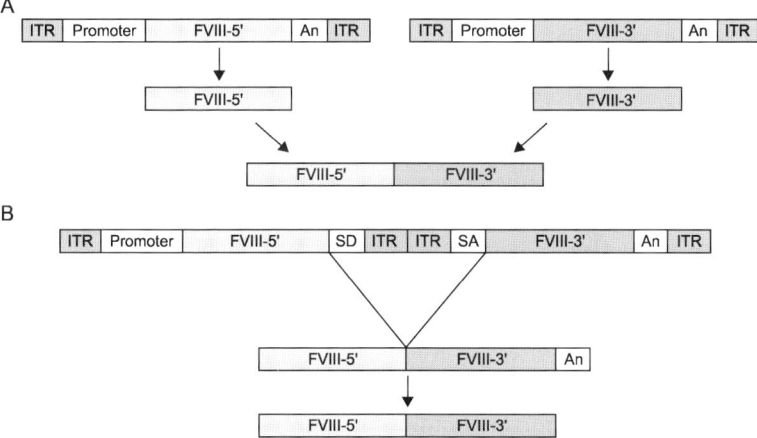

Fig. 3.2. Schematic of recombinant AAV dual-vector strategies for gene therapy for hemophilia A. (A) depicts the individual expression of the 5′ and 3′ ends of the factor VIII protein followed by heterodimerization to generate a function protein. (B) shows concatamerized 5′ and 3′ vectors in the correct head-to-tail orientation. Splicing of vectors to remove the intron and inverted terminal repeats leads to expression of the entire factor VIII protein from a single mature mRNA transcript. ITR, AAV inverted terminal repeat; SD, splice donor; SA, splice acceptor; An, poly A.

In recent years, there has been a great deal of interest in alternate rAAV serotype vectors as substitutes for the well-characterized and often-used AAV serotype 2 vectors. To date, at least 40 non-redundant clones of AAV have been identified, eight of which have been developed as rAAV vectors (Chiorini et al., 1997, 1999; Rutledge et al., 1998; Gao et al., 2002; Rabinowitz et al., 2002). Most pre-clinical studies for bleeding disorders as well as the two AAV clinical trials for hemophilia B have been based on AAV2 vectors. Recent work has demonstrated a distinction in tissue type expression by the different serotype vectors. Several groups have shown AAV1 vectors to transduce skeletal muscle even better than AAV2 vectors (Fraites et al., 2002; Gao et al., 2002; Rabinowitz et al., 2002). The higher transduction efficiencies seen with AAV1 vectors may be due to a lack of muscle fiber-type restriction for infection. Using AAV1 vectors carrying the canine factor IX gene in hemophilia B mice, Chao et al. reported supraphysiologic levels of factor IX levels 3–5 weeks post-intramuscular administration. By contrast, AAV2 vectors resulted in only 2% normal levels of canine factor IX (Chao et al., 2000a, 2001). Other studies have demonstrated that AAV5 delivery results in significantly higher liver transduction efficiencies as compared to type 2. Mingozzi et al. demonstrated approximately 3-fold higher levels of hepatocyte transduction, resulting in 3- to 10-fold higher levels of human factor IX levels in C57BL/6 mice as compared to AAV2-mediated expression levels of approximately 200 ng/ml (Mingozzi et al., 2002). More recent work by several groups suggests that AAV serotype 8 vectors may be even more efficient than serotype 5 vectors in transducing liver (Gao et al., 2002). Sarkar et al. and Wang et al. noted physiologic and supraphysiologic levels of factor VIII and factor IX in AAV serotype 8 vector-treated hemophilia A and B mice, respectively, as well as partial correction in the corresponding dog models of hemophilia (Sarkar et al., 2003; Wang et al., 2003). It is important to note that these AAV serotype-dependent transduction efficiencies have been determined in vitro and in murine and canine tissues in vivo; whether the same specificity occurs in human applications remain to be seen. Although these newer serotype vectors may transduce particular tissue types more efficiently, the route or method of vector administration may also play a role in the efficient transduction of the target cell type. Methods used to administer AAV vectors for hemophilia gene therapy have included more simple intramuscular or intravenous injections to more complex methods of intrahepatic arterial infusion or isolated limb-perfusion (Herzog et al., 2001; Arruda et al., 2002; Mount et al., 2002; Nathwani et al., 2002; Mah et al., 2003; Manno et al., 2003). In the context of bleeding disorders, a vector that can be administered via simple intravenous injection, target the desired tissue type, while avoiding transduction of non-target cells, would be ideal. Not only would the method of administration be minimally traumatic to the patient, but such a targeting vector would also enable treatment to be available to a larger population of patients, especially those who do not have access to the facilities required for more complicated means of vector delivery, such as catheterization to achieve portal administration. A recent development to enhance targeting to specific cell types involves the inclusion of targeting moieties in the AAV capsid itself. While targeting AAV vector technology has not yet been specifically applied to the development of vectors for bleeding disorders, targeting vectors have the potential to be minimally pathogenic and more efficient than currently available vector systems (Yang et al., 1998; Girod et al., 1999; Wu et al., 2000; Shi et al., 2001; Nicklin and Baker, 2002; Ponnazhagan et al., 2002; Shi and Bartlett, 2003).

3.5. Immunological considerations

Since the mid 1960s, with the discovery of cryoprecipitate by Judith Graham Pool, there has been a growing understanding of the biology associated with protein replacement therapies (Pool and Shannon, 1965). A serious recurring obstacle to protein replacement therapies is the development of inhibitory antibodies in some patients to the therapeutic protein. Inhibitory antibodies can render a patient completely refractory to standard modes of protein replacement therapies. According to the World Federation of Hemophilia, between 20 and 35% of patients with severe or moderately severe hemophilia A develop inhibitors. For this reason, gene therapy strategies are being developed with the added goal of avoiding inhibitor formation. In pre-clinical animal model studies to this point, the development of inhibitory antibodies resulting from gene therapy has not been well characterized or understood because inhibitor formation has been inconsistent. These inconsistencies have been attributed to the various differences between these studies, and several factors have been implicated to play a role the potential for inhibitor formation. Among these potential factors are the immunogenicity of the vector system; the route of administration; the age at which therapy is given; the variable levels of expression; the target organ or tissue; the cell-type specificity of expression; the form of the transgene used; and the nature of the genetic mutation that lead to the diseased state (Snyder et al., 1997; Vandendriessche et al., 1999; Balague et al., 2000; Xu et al., 2001, 2003; Herzog et al., 2002; Reddy et al., 2002; Mah et al., 2003; Mingozzi et al., 2003).

Several methods have been proposed to minimize the potential for inhibitor formation. Since the age at which therapy is initiated may play a role in inhibitor formation, several studies have used neonatal animals, in the hope that the immature immune system will not recognize the vector-derived protein as foreign (Xu et al., 2001, 2003; Mah et al., 2003). Other studies have gone one step earlier, delivering vectors in utero (Lipshutz et al., 1999; Schneider et al., 1999). The use of improved therapeutic transgenes may also improve the likelihood of success (Toole et al., 1986; Lynch et al., 1993; Pittman et al., 1993; Hoeben et al., 1995; Kaufman and Pipe, 1998, 1999; Miao et al., 2000; Pipe et al., 2001; Saenko et al., 2003). Factor VIII genes that result in protein devoid of antigenic epitopes or with longer half-lives are under development. Finally, several groups have demonstrated an induced immune tolerance to gene therapy vector-derived expression (Chao and Walsh, 2001; Herzog et al., 2002; Mingozzi et al., 2003). While the mechanism of immune tolerance is still not well understood and warrants further investigation, work by Mingozzi et al. demonstrated immune tolerance in several different strains of adult immunocompetent mice treated with an AAV2-human factor IX vector (Mingozzi et al., 2003). It was suggested that factor IX tolerance was induced by targeting the vector to liver, both by the route of administration and the promoter element used, and by the high level of transgene expression. In response to the subset of hemophilic patients that have high inhibitor titers and are therefore resistant to current protein therapy treatments, other proteins such as porcine factor VIII and factor VIIa have been produced (Brettler et al., 1989; Lusher, 1996; Scandella et al., 2001; Hakeos et al., 2002). Porcine factor VIII is resistant to antihuman factor VIII antibodies and factor VIIa can activate the extrinsic coagulation pathway, thereby circumventing the need for factor VIII or IX. Since factor VIII or factor IX gene therapy vectors may also be rendered useless in subjects with high inhibitor titers, gene therapy vectors expressing an inducible factor VIIa or antigenically modified protein are also currently under development (Margaritis et al., 2003).

For a number of historic and public health reasons, gene therapy approaches for bleeding disorders were among the first attempted using these new technologies. Like the field of gene therapy in general, therapeutic strategies for hemophilia have undergone significant changes and improvements since the early 1990s when these studies began in earnest. Ongoing pre-clinical and clinical trials seek to demonstrate the utility of these approaches while gaining valuable understandings of the etiology of bleeding disorders and the interaction between therapeutic vectors and human subjects. The advent of more precisely targeted vector systems, combined with tissue-specific promoters and a better understanding of both vector- and transgene-related immunology, promises to provide patients with safer, more effective therapies, offering the potential for long-term correction and enhancing their overall quality of life.

3.6. Laboratory protocols

Several protocols are routinely used in assessing the success of gene therapy vector treatment in the animal models of hemophilia and are described below.

3.6.1. Detection of biologically active factor VIII

Levels of biologically active factor VIII can be measured using the Coatest chromogenic kit assay (Chromogenix, Milan, Italy). Plasma samples are obtained by bleeding from anesthetized animals followed by the immediate addition of sodium citrate to a final concentration of 0.38% (wt/vol). Samples are then centrifuged at 2000 × g for 10 min at 25°C. The plasma fraction is collected and frozen immediately. Samples are thawed quickly at 37°C immediately prior to testing. Serial dilutions of pooled normal animal plasma diluted in pooled hemophilic animal plasma are used to derive the standard curve (Burton et al., 1999; Vandendriessche et al., 1999; Balague et al., 2000; Sarkar et al., 2000; Reddy et al., 2002; Mah et al., 2003).

3.6.2. Detection of the factor VIII or IX antigen

A standard double-sandwich enzyme-linked immunosorbent assay (ELISA) or Western blot analysis is used. As the concentration of factor is very low in normal plasma (approximately 0.1 mg/l for factor VIII), it is necessary to subject plasma samples to cryoprecipitation in order to concentrate the sample, prior to Western blot analysis. The cryoprecipitation protocol described by Bi et al. is as follows (Bi et al., 1996; Sarkar et al., 2000; Mah et al., 2003). Plasma samples are collected as described above for the Coatest assay. Plasma is then frozen at –80°C overnight. Frozen samples are then subject to centrifugation at 7000 × g for 20 min at 4°C. The precipitate is washed with

300 μl ice-cold 20 mM Tris–HCl, pH 7.0, and centrifuged again at 7000 × g for 20 min. The resulting precipitate is resuspended in 30 μl 20 mM Tris–HCl, pH 7.0. Glycine buffer (75 μl) (2.8 M glycine, 0.3 M $NaCl_2$, 25 mM Tris, pH 6.8) is added and the sample is incubated at room temperature for 30 min. The sample is centrifuged at 7000 × g for 10 min to remove any insoluble precipitate. Supernatant (25 μl) can be used for Western blotting.

3.6.3. Activated partial thromboplastin time (aPTT)

Equal amounts (50 μl) of plasma sample with citrate buffer and aPTT reagent are mixed and incubated at 37 °C for exactly 3 min. $CaCl_2$ (50 μl; 25 mM) is added to the sample and the clotting time is measured with a fibrometer (Herzog et al., 1999; Chao et al., 2001; Fields et al., 2001).

3.6.4. Bethesda assay

The Bethesda assay is used to detect levels of inhibitory antibodies. One Bethesda unit is the reciprocal of the dilution of test plasma at which 50% of normal activity is inhibited. Briefly, the test plasma sample is mixed with an equal volume of pooled normal plasma and incubated at 37 °C for 2 h. Residual activity is measured by aPTT as described above (Herzog et al., 1999; Chao et al., 2001; Fields et al., 2001; Gallo-Penn et al., 2001).

References

Andrews, J. L., Kadan, M. J., Gorziglia, M. I., Kaleko, M. and Connelly, S. (2001). Generation and characterization of E1/E2a/E3/E4-deficient adenoviral vectors encoding human factor VIII. Mol. Ther. 3, 329–336.

Arruda, V. R., Stedman, H., Nichols, T. C., Haskind, M., Schuettrumpf, J., Herzog, R. W., Liu, Y. L., Wagner, K., Couto, L. and High, K. A. (2002).

Sustained correction of hemophilia B phenotype following intravascular delivery of AAV vector to skeletal muscle. Mol. Ther. 5, S157.

Balague, C., Zhou, J., Dai, Y., Alemany, R., Josephs, S. F., Andreason, G., Hariharan, M., Sethi, E., Prokopenko, E., Jan, H. Y., Lou, Y. C., Hubert-Leslie, D., Ruiz, L. and Zhang, W. W. (2000). Sustained high-level expression of full-length human factor VIII and restoration of clotting activity in hemophilic mice using a minimal adenovirus vector. Blood 95, 820–828.

Bi, L., Lawler, A. M., Antonarakis, S. E., High, K. A., Gearhart, J. D. and Kazazian, H. H., Jr. (1995). Targeted disruption of the mouse factor VIII gene produces a model of haemophilia A. Nat. Genet. 10, 119–121.

Bi, L., Sarkar, R., Naas, T., Lawler, A. M., Pain, J., Shumaker, S. L., Bedian, V. and Kazazian, H. H., Jr. (1996). Further characterization of factor VIII-deficient mice created by gene targeting: RNA and protein studies. Blood 88, 3446–3450.

Brettler, D. B., Forsberg, A. D., Levine, P. H., Aledort, L. M., Hilgartner, M. W., Kasper, C. K., Lusher, J. M., McMillan, C. and Roberts, H. (1989). The use of porcine factor VIII concentrate (Hyate:C) in the treatment of patients with inhibitor antibodies to factor VIII. A multicenter US experience. Arch. Intern. Med. 149, 1381–1385.

Bristol, J. A., Gallo-Penn, A., Andrews, J., Idamakanti, N., Kaleko, M. and Connelly, S. (2001). Adenovirus-mediated factor VIII gene expression results in attenuated anti-factor VIII-specific immunity in hemophilia A mice compared with factor VIII protein infusion. Hum. Gene Ther. 12, 1651–1661.

Burton, M., Nakai, H., Colosi, P., Cunningham, J., Mitchell, R. and Couto, L. (1999). Coexpression of factor VIII heavy and light chain adeno-associated viral vectors produces biologically active protein. Proc. Natl. Acad. Sci. USA 96, 12725–12730.

Chao, H. and Walsh, C. E. (2001). Induction of tolerance to human factor VIII in mice. Blood 97, 3311–3312.

Chao, H., Mao, L., Bruce, A. T. and Walsh, C. E. (2000a). Sustained expression of human factor VIII in mice using a parvovirus-based vector. Blood 95, 1594–1599.

Chao, H., Liu, Y., Rabinowitz, J., Li, C., Samulski, R. J. and Walsh, C. E. (2000b). Several log increase in therapeutic transgene delivery by distinct adeno-associated viral serotype vectors. Mol. Ther. 2, 619–623.

Chao, H., Sun, L., Bruce, A., Xiao, X. and Walsh, C. E. (2002). Expression of human factor VIII by splicing between dimerized AAV vectors. Mol. Ther. 5, 716–722.

Chao, H., Monahan, P. E., Liu, Y., Samulski, R. J. and Walsh, C. E. (2001). Sustained and complete phenotype correction of hemophilia B mice following intramuscular injection of AAV1 serotype vectors. Mol. Ther. 4, 217–222.

Chiorini, J. A., Kim, F., Yang, L. and Kotin, R. M. (1999). Cloning and characterization of adeno-associated virus type 5. J. Virol. 73, 1309–1319.

Chiorini, J. A., Yang, L., Liu, Y., Safer, B. and Kotin, R. M. (1997). Cloning of adeno-associated virus type 4 (AAV4) and generation of recombinant AAV4 particles. J. Virol. 71, 6823–6833.

Chuah, M. K., Van Damme, A., Zwinnen, H., Goovaerts, I., Vanslembrouck, V., Collen, D. and Vandendriessche, T. (2000). Long-term persistence of human bone marrow stromal cells transduced with factor VIII-retroviral vectors and transient production of therapeutic levels of human factor VIII in nonmyeloablated immunodeficient mice. Hum. Gene Ther. 11, 729–738.

Chuah, M. K., Schiedner, G., Thorrez, L., Brown, B., Johnston, M., Gillijns, V., Hertel, S., Van Rooijen, N., Lillicrap, D., Collen, D., Vandendriessche, T. and Kochanek, S. (2003). Therapeutic factor VIII levels and negligible toxicity in mouse and dog models of hemophilia A following gene therapy with high-capacity adenoviral vectors. Blood 101, 1734–1743.

Connelly, S. and Kaleko, M. (1998). Haemophilia A gene therapy. Haemophilia 4, 380–388.

Connelly, S., Andrews, J. L., Gallo-Penn, A. M., Tagliavacca, L., Kaufman, R. J. and Kaleko, M. (1999). Evaluation of an adenoviral vector encoding full-length human factor VIII in hemophiliac mice. J. Thromb. Haemost. 81, 234–239.

Duan, D., Yue, Y., Yan, Z. and Engelhardt, J. F. (2003). Trans-splicing vectors expand the packaging limits of adeno-associated virus for gene therapy applications. Methods Mol. Med. 76, 287–307.

Dwarki, V. J., Belloni, P., Nijjar, T., Smith, J., Couto, L., Rabier, M., Clift, S., Berns, A. and Cohen, L. K. (1995). Gene therapy for hemophilia A: Production of therapeutic levels of human factor VIII in vivo in mice. Proc. Natl. Acad. Sci. USA 92, 1023–1027.

Ehrhardt, A. and Kay, M. A. (2002). A new adenoviral helper-dependent vector results in long-term therapeutic levels of human coagulation factor IX at low doses in vivo. Blood 99, 3923–3930.

Evans, J. P., Brinkhous, K. M., Brayer, G. D., Reisner, H. M. and High, K. A. (1989). Canine hemophilia B resulting from a point mutation with unusual consequences. Proc. Natl. Acad. Sci. USA 86, 10095–10099.

Fields, P. A., Arruda, V. R., Armstrong, E., Chu, K., Mingozzi, F., Hagstrom, J. N., Herzog, R. W. and High, K. A. (2001). Risk and prevention of anti-factor IX formation in AAV-mediated gene transfer in the context of a large deletion of F9. Mol. Ther. 4, 201–210.

Fraites, T. J., Jr., Schleissing, M. R., Shanely, R. A., Walter, G. A., Cloutier, D. A., Zolotukhin, I., Pauly, D. F., Raben, N., Plotz, P. H., Powers, S. K., Kessler, P. D. and Byrne, B. J. (2002). Correction of the enzymatic and functional deficits in a model of Pompe disease using adeno-associated virus vectors. Mol. Ther. 5, 571–578.

Gallo-Penn, A. M., Shirley, P. S., Andrews, J. L., Tinlin, S., Webster, S., Cameron, C., Hough, C., Notley, C., Lillicrap, D., Kaleko, M. and Connelly, S. (2001). Systemic delivery of an adenoviral vector encoding canine factor VIII results in short-term phenotypic correction, inhibitor development, and biphasic liver toxicity in hemophilia A dogs. Blood 97, 107–113.

Gao, G. P., Alvira, M. R., Wang, L., Calcedo, R., Johnston, J. and Wilson, J. M. (2002). Novel adeno-associated viruses from rhesus monkeys as vectors for human gene therapy. Proc. Natl. Acad. Sci. USA 99, 11854–11859.

Garcia-Martin, C., Chuah, M. K., Van Damme, A., Robinson, K. E., Vanzieleghem, B., Saint-Remy, J. M., Gallardo, D., Ofosu, F. A., Vandendriessche, T. and Hortelano, G. (2002). Therapeutic levels of human factor VIII in mice implanted with encapsulated cells: Potential for gene therapy of haemophilia A. J. Gene Med. 4, 215–223.

Girod, A., Ried, M., Wobus, C., Lahm, H., Leike, K., Kleinschmidt, J., Deleage, G. and Hallek, M. (1999). Genetic capsid modifications allow efficient re-targeting of adeno-associated virus type 2. Nat. Med. 5, 1438.

Gnatenko, D. V., Saenko, E. L., Jesty, J., Cao, L. X., Hearing, P. and Bahou, W. F. (1999). Human factor VIII can be packaged and functionally expressed in an adeno-associated virus background: Applicability to haemophilia A gene therapy. Br. J. Haematol. 104, 27–36.

Hakeos, W. H., Miao, H., Sirachainan, N., Kemball-Cook, G., Saenko, E. L., Kaufman, R. J. and Pipe, S. W. (2002). Hemophilia A mutations within the factor VIII A2-A3 subunit interface destabilize factor VIIIa and cause one-stage/two-stage activity discrepancy. J. Thromb. Haemost. 88, 781–787.

Hao, Q. L., Malik, P., Salazar, R., Tang, H., Gordon, E. M. and Kohn, D. B. (1995). Expression of biologically active human factor IX in human hematopoietic cells after retroviral vector-mediated gene transduction. Hum. Gene Ther. 6, 873–880.

Herzog, R. W., Yang, E. Y., Couto, L. B., Hagstrom, J. N., Elwell, D., Fields, P. A., Burton, M., Bellinger, D. A., Read, M. S., Brinkhous, K. M., Podsakoff, G. M., Nichols, T. C., Kurtzman, G. J. and High, K. A. (1999). Long-term correction of canine hemophilia B by gene transfer of blood coagulation factor IX mediated by adeno-associated viral vector. Nat. Med. 5, 56–63.

Herzog, R. W., Mount, J. D., Arruda, V. R., High, K. A. and Lothrop, C. D., Jr. (2001). Muscle-directed gene transfer and transient immune suppression result in sustained partial correction of canine hemophilia B caused by a null mutation. Mol. Ther. *4*, 192–200.

Herzog, R. W., Fields, P. A., Arruda, V. R., Brubaker, J. O., Armstrong, E., McClintock, D., Bellinger, D. A., Couto, L. B., Nichols, T. C. and High, K. A. (2002). Influence of vector dose on factor IX-specific T and B cell responses in muscle-directed gene therapy. Hum. Gene Ther. *13*, 1281–1291.

High, K. A. (2003). Gene transfer as an approach to treating hemophilia. Semin. Thromb. Hemost. *29*, 107–120.

Hoeben, R. C., Fallaux, F. J., Cramer, S. J., van den Wollenberg, D. J., van Ormondt, H., Briet, E. and Van Der Eb, A. J. (1995). Expression of the blood-clotting factor-VIII cDNA is repressed by a transcriptional silencer located in its coding region. Blood *85*, 2447–2454.

Hortelano, G., Wang, L., Xu, N. and Ofosu, F. A. (2001). Sustained and therapeutic delivery of factor IX in nude haemophilia B mice by encapsulated C2C12 myoblasts: Concurrent tumourigenesis. Haemophilia *7*, 207–214.

Hough, C., Kamisue, S., Cameron, C., Notley, C., Tinlin, S., Giles, A. and Lillicrap, D. (2002). Aberrant splicing and premature termination of transcription of the FVIII gene as a cause of severe canine hemophilia A: Similarities with the intron 22 inversion mutation in human hemophilia. Thromb. Haemost. *87*, 659–665.

Hoyer, L. W. (1994). Hemophilia A. N. Engl. J. Med. *330*, 38–47.

Kaufman, R. J. and Pipe, S. W. (1998). Can we improve on nature? "Super molecules" of factor VIII. Haemophilia *4*, 370–379.

Kaufman, R. J. and Pipe, S. W. (1999). Regulation of factor VIII expression and activity by von Willebrand factor. Thromb. Haemost. *82*, 201–208.

Kay, M. A., Manno, C. S., Ragni, M. V., Larson, P. J., Couto, L. B., McClelland, A., Glader, B., Chew, A. J., Tai, S. J., Herzog, R. W., Arruda, V., Johnson, F., Scallan, C., Skarsgard, E., Flake, A. W. and High, K. A. (2000). Evidence for gene transfer and expression of factor IX in haemophilia B patients treated with an AAV vector. Nat. Genet. *24*, 257–261.

Kelley, K., Verma, I. and Pierce, G. F. (2002). Gene therapy: Reality or myth for the global bleeding disorders community? Haemophilia *8*, 261–267.

Kootstra, N. A., Matsumura, R. and Verma, I. M. (2003). Efficient production of human FVIII in hemophilic mice using lentiviral vectors. Mol. Ther. *7*, 623–631.

Krebsbach, P. H., Zhang, K., Malik, A. K. and Kurachi, K. (2003). Bone marrow stromal cells as a genetic platform for systemic delivery of therapeutic proteins in vivo: Human factor IX model. J. Gene Med. *5*, 11–17.

Lin, H. F., Maeda, N., Smithies, O., Straight, D. L. and Stafford, D. W. (1997). A coagulation factor IX-deficient mouse model for human hemophilia B. Blood 90, 3962–3966.

Lin, Y., Chang, L., Solovey, A., Healey, J. F., Lollar, P. and Hebbel, R. P. (2002). Use of blood outgrowth endothelial cells for gene therapy for hemophilia A. Blood 99, 457–462.

Lipshutz, G. S., Sarkar, R., Flebbe-Rehwaldt, L., Kazazian, H. and Gaensler, K. M. (1999). Short-term correction of factor VIII deficiency in a murine model of hemophilia A after delivery of adenovirus murine factor VIII in utero. Proc. Natl. Acad. Sci. USA 96, 13324–13329.

Lozier, J. N., Dutra, A., Pak, E., Zhou, N., Zheng, Z., Nichols, T. C., Bellinger, D. A., Read, M. and Morgan, R. A. (2002). The Chapel Hill hemophilia A dog colony exhibits a factor VIII gene inversion. Proc. Natl. Acad. Sci. USA 99, 12991–12996.

Lusher, J. M. (1996). Recombinant factor VIIa (NovoSeven) in the treatment of internal bleeding in patients with factor VIII and IX inhibitors. Haemostasis 26(Suppl. 1), 124–130.

Lynch, C. M., Israel, D. I., Kaufman, R. J. and Miller, A. D. (1993). Sequences in the coding region of clotting factor VIII act as dominant inhibitors of RNA accumulation and protein production. Hum. Gene Ther. 4, 259–272.

Mah, C., Sarkar, R., Zolotukhin, I., Schleissing, M., Xiao, X., Kazazian, H. H. and Byrne, B. J. (2003). Dual vectors expressing murine factor VIII result in sustained correction of hemophilia A mice. Hum. Gene Ther. 14, 143–152.

Manno, C. S., Chew, A. J., Hutchison, S., Larson, P. J., Herzog, R. W., Arruda, V. R., Tai, S. J., Ragni, M. V., Thompson, A., Ozelo, M., Couto, L. B., Leonard, D. G., Johnson, F. A., McClelland, A., Scallan, C., Skarsgard, E., Flake, A. W., Kay, M. A., High, K. A. and Glader, B. (2003). AAV-mediated factor IX gene transfer to skeletal muscle in patients with severe hemophilia B. Blood 101, 2963–2972.

Margaritis, P., Arruda, V. R. and High, K. A. (2003). Gene therapy for hemophilia with inhibitors using secreted activated FVII (FVIIa). Mol. Ther. 7, S27.

Miao, C. H., Thompson, A. R., Loeb, K. and Ye, X. (2001). Long-term and therapeutic-level hepatic gene expression of human factor IX after naked plasmid transfer in vivo. Mol. Ther. 3, 947–957.

Miao, C. H., Ohashi, K., Patijn, G. A., Meuse, L., Ye, X., Thompson, A. R. and Kay, M. A. (2000). Inclusion of the hepatic locus control region, an intron, and untranslated region increases and stabilizes hepatic factor IX gene expression in vivo but not in vitro. Mol. Ther. 1, 522–532.

Mingozzi, F., Schuttrumpf, J., Arruda, V. R., Liu, Y., Liu, Y. L., High, K. A., Xiao, W. and Herzog, R. W. (2002). Improved hepatic gene transfer by using an adeno-associated virus serotype 5 vector. J. Virol. 76, 10497–10502.

Mingozzi, F., Liu, Y. L., Dobrzynski, E., Kaufhold, A., Liu, J. H., Wang, Y., Arruda, V. R., High, K. A. and Herzog, R. W. (2003). Induction of immune tolerance to coagulation factor IX antigen by in vivo hepatic gene transfer. J. Clin. Invest. 111, 1347–1356.

Mount, J. D., Herzog, R. W., Tillson, D. M., Goodman, S. A., Robinson, N., McCleland, M. L., Bellinger, D., Nichols, T. C., Arruda, V. R., Lothrop, C. D., Jr. and High, K. A. (2002). Sustained phenotypic correction of hemophilia B dogs with a factor IX null mutation by liver-directed gene therapy. Blood 99, 2670–2676.

Nathwani, A. C., Davidoff, A. M., Hanawa, H., Hu, Y., Hoffer, F. A., Nikanorov, A., Slaughter, C., Ng, C. Y., Zhou, J., Lozier, J. N., Mandrell, T. D., Vanin, E. F. and Nienhuis, A. W. (2002). Sustained high-level expression of human factor IX (hFIX) after liver-targeted delivery of recombinant adeno-associated virus encoding the hFIX gene in rhesus macaques. Blood 100, 1662–1669.

Nicklin, S. A. and Baker, A. H. (2002). Tropism-modified adenoviral and adeno-associated viral vectors for gene therapy. Curr. Gene Ther. 2, 273–293.

Park, F., Ohashi, K. and Kay, M. A. (2000). Therapeutic levels of human factor VIII and IX using HIV-1-based lentiviral vectors in mouse liver. Blood 96, 1173–1176.

Pipe, S. W., Saenko, E. L., Eickhorst, A. N., Kemball-Cook, G. and Kaufman, R. J. (2001). Hemophilia A mutations associated with 1-stage/2-stage activity discrepancy disrupt protein–protein interactions within the triplicated A domains of thrombin-activated factor VIIIa. Blood 97, 685–691.

Pittman, D. D., Alderman, E. M., Tomkinson, K. N., Wang, J. H., Giles, A. R. and Kaufman, R. J. (1993). Biochemical, immunological, and in vivo functional characterization of B-domain-deleted factor VIII. Blood 81, 2925–2935.

Ponnazhagan, S., Mahendra, G., Kumar, S., Thompson, J. A. and Castillas, M., Jr. (2002). Conjugate-based targeting of recombinant adeno-associated virus type 2 vectors by using avidin-linked ligands. J. Virol. 76, 12900–12907.

Pool, J. G. and Shannon, A. E. (1965). Production of high-potency concentrates of antihemophilic globulin in a closed-bag system. N. Engl. J. Med. 273, 1443–1447.

Powell, J. S., Ragni, M. V., White, G. C., Lusher, J. M., Hillman-Wiseman, C., Moon, T. E., Cole, V., Ramanathan-Girish, S., Roehl, H., Sajjadi, N., Jolly, D. J. and Hurst, D. (2003). Phase I trial of FVIII gene transfer for severe hemophilia A using a retroviral construct administered by peripheral intravenous infusion. Blood *102*, 2038–2045.

Rabinowitz, J. E., Rolling, F., Li, C., Conrath, H., Xiao, W., Xiao, X. and Samulski, R. J. (2002). Cross-packaging of a single adeno-associated virus (AAV) type 2 vector genome into multiple AAV serotypes enables transduction with broad specificity. J. Virol. *76*, 791–801.

Reddy, P. S., Sakhuja, K., Ganesh, S., Yang, L., Kayda, D., Brann, T., Pattison, S., Golightly, D., Idamakanti, N., Pinkstaff, A., Kaloss, M., Barjot, C., Chamberlain, J. S., Kaleko, M. and Connelly, S. (2002). Sustained human factor VIII expression in hemophilia A mice following systemic delivery of a gutless adenoviral vector. Mol. Ther. *5*, 63–73.

Roth, D. A., Tawa, N. E., Jr., O' Brien, J. M., Treco, D. A. and Selden, R. F. (2001). Nonviral transfer of the gene encoding coagulation factor VIII in patients with severe hemophilia A. N. Engl. J. Med. *344*, 1735–1742.

Rutledge, E. A., Halbert, C. L. and Russell, D. W. (1998). Infectious clones and vectors derived from adeno-associated virus (AAV) serotypes other than AAV type 2. J. Virol. *72*, 309–319.

Saenko, E. L., Ananyeva, N. M., Moayeri, M., Ramezani, A. and Hawley, R. G. (2003). Development of improved factor VIII molecules and new gene transfer approaches for hemophilia A. Curr. Gene Ther. *3*, 27–41.

Sarkar, R., Gao, G. P., Chirmule, N., Tazelaar, J. and Kazazian, H. H., Jr. (2000). Partial correction of murine hemophilia A with neo-antigenic murine factor VIII. Hum. Gene Ther. *11*, 881–894.

Sarkar, R., Tetreault, R., Gao, G. P., Wang, L., Bell, P., Chandler, R., Bellinger, D., Nichols, T. C., Wilson, J. M. and Kazazian, H. H. (2003). AAV 8 serotype: Total correction of hemophilia A mice and partial correction in dogs with canine FVIII. Mol. Ther. *7*, S26.

Scallan, C. D., Lillicrap, D., Jiang, H., Qian, X., Patarroyo-White, S. L., Parker, A. E., Liu, T., Vargas, J., Nagy, D., Powell, S. K., Wright, J. F., Turner, P. V., Tinlin, S. J., Webster, S. E., McClelland, A. and Couto, L. B. (2003). Sustained phenotypic correction of canine hemophilia A using an adeno-associated viral vector. Blood *102*, 2031–2037.

Scandella, D. H., Nakai, H., Felch, M., Mondorf, W., Scharrer, I., Hoyer, L. W. and Saenko, E. L. (2001). In hemophilia A and autoantibody inhibitor patients: The factor VIII A2 domain and light chain are most immunogenic. Thromb. Res. *101*, 377–385.

Schneider, H., Adebakin, S., Themis, M., Cook, T., Douar, A. M., Pavirani, A. and Coutelle, C. (1999). Therapeutic plasma concentrations of human factor IX in mice after gene delivery into the amniotic cavity: A model for the prenatal treatment of haemophilia B. J. Gene Med. *1*, 424–432.

Shi, W. and Bartlett, J. S. (2003). RGD inclusion in VP3 provides adeno-associated virus type 2 (AAV2)-based vectors with a heparan sulfate-independent cell entry mechanism. Mol. Ther. *7*, 515–525.

Shi, W., Arnold, G. S. and Bartlett, J. S. (2001). Insertional mutagenesis of the adeno-associated virus type 2 (AAV2) capsid gene and generation of AAV2 vectors targeted to alternative cell-surface receptors. Hum. Gene Ther. *12*, 1697–1711.

Snyder, R. O., Miao, C. H., Patijn, G. A., Spratt, S. K., Danos, O., Nagy, D., Gown, A. M., Winther, B., Meuse, L., Cohen, L. K., Thompson, A. R. and Kay, M. A. (1997). Persistent and therapeutic concentrations of human factor IX in mice after hepatic gene transfer of recombinant AAV vectors. Nat. Genet. *16*, 270–276.

Stein, C. S., Kang, Y., Sauter, S. L., Townsend, K., Staber, P., Derksen, T. A., Martins, I., Qian, J., Davidson, B. L. and McCray, P. B., Jr. (2001). In vivo treatment of hemophilia A and mucopolysaccharidosis type VII using nonprimate lentiviral vectors. Mol. Ther. *3*, 850–856.

Sun, L., Li, J. and Xiao, X. (2000). Overcoming adeno-associated virus vector size limitation through viral DNA heterodimerization. Nat. Med. *6*, 599–602.

Toole, J. J., Pittman, D. D., Orr, E. C., Murtha, P., Wasley, L. C. and Kaufman, R. J. (1986). A large region (approximately equal to 95 kDa) of human factor VIII is dispensable for in vitro procoagulant activity. Proc. Natl. Acad. Sci. USA *83*, 5939–5942.

Van Damme, A., Chuah, M. K., Dell'accio, F., De Bari, C., Luyten, F., Collen, D. and Vandendriessche, T. (2003). Bone marrow mesenchymal cells for haemophilia A gene therapy using retroviral vectors with modified long-terminal repeats. Haemophilia *9*, 345.

Vandendriessche, T., Vanslembrouck, V., Goovaerts, I., Zwinnen, H., Vanderhaeghen, M. L., Collen, D. and Chuah, M. K. (1999). Long-term expression of human coagulation factor VIII and correction of hemophilia A after in vivo retroviral gene transfer in factor VIII-deficient mice. Proc. Natl. Acad. Sci. USA *96*, 10379–10384.

Wang, L., Gao, G. P., Bell, P., Sanmiguel, J., Li, Y., Nichols, T. C., Verma, I. M. and Wilson, J. M. (2003). Gene therapy for hemophilia B by novel serotypes of AAV vectors. Mol. Ther. *7*, S26.

Wu, P., Xiao, W., Conlon, T., Hughes, J., Agbandje-McKenna, M., Ferkol, T., Flotte, T. and Muzyczka, N. (2000). Mutational analysis of the adeno-associated virus type 2 (AAV2) capsid gene and construction of AAV2 vectors with altered tropism. J. Virol. *74*, 8635–8647.

Xu, L., Gao, C., Sands, M. S., Cai, S. R., Nichols, T. C., Bellinger, D. A., Raymer, R. A., McCorquodale, S. and Ponder, K. P. (2003). Neonatal or hepatocyte growth factor-potentiated adult gene therapy with a retroviral vector results in therapeutic levels of canine factor IX for hemophilia B. Blood *101*, 3924–3932.

Xu, L., Daly, T., Gao, C., Flotte, T. R., Song, S., Byrne, B. J., Sands, M. S. and Parker, P. K. (2001). CMV-beta-actin promoter directs higher expression from an adeno-associated viral vector in the liver than the cytomegalovirus or elongation factor 1 alpha promoter and results in therapeutic levels of human factor X in mice. Hum. Gene Ther. *12*, 563–573.

Yang, Q., Mamounas, M., Yu, G., Kennedy, S., Leaker, B., Merson, J., Wong-Staal, F., Yu, M. and Barber, J. R. (1998). Development of novel cell surface CD34-targeted recombinant adenoassociated virus vectors for gene therapy. Hum. Gene Ther. *9*, 1929–1937.

Yant, S. R., Meuse, L., Chiu, W., Ivics, Z., Izsvak, Z. and Kay, M. A. (2000). Somatic integration and long-term transgene expression in normal and haemophilic mice using a DNA transposon system. Nat. Genet. *25*, 35–41.

Laboratory Techniques in Biochemistry and Molecular Biology, Volume 31
Adeno-Associated Viral Vectors for Gene Therapy
T. R. Flotte and K. I. Berns (Editors)

CHAPTER 4

Recombinant AAV vectors for gene transfer to the lung: a compartmental approach

Terence R. Flotte

Departments of Pediatrics and Molecular Genetics and Microbiology, Powell Gene Therapy Center and Genetics Institute, University of Florida, Box 100296, 1600 SW Archer Rd., Gainesville, FL 32610-0296, USA

4.1. Introduction

Ironically, although the lung was the first site of delivery for which in vivo gene transfer with rAAV was demonstrated (Flotte et al., 1993a) and the first for which human trials were initiated (Flotte et al., 1996), it still remains a difficult target for gene transfer. The examination of barriers to efficient rAAV2-mediated gene transfer in the airways in cystic fibrosis (CF) has helped identify key steps for AAV entry, nuclear transport, genome processing, and transcription activation. It has further led to innovations in vector design to help overcome these barriers. Meanwhile, rAAV mediated gene transfer may also have a place for gene transfer in other cell types within the lung, such as alveoli and vascular structures, which may be targeted for gene therapy of other diseases. Each of these different cell types within the lung may behave differently with regard to rAAV transduction.

© 2005 Elsevier B.V. All rights reserved
DOI: 10.1016/S0075-7535(05)31004-7

4.2. Genes, targets and vectors for the lung

In most cases, the first task in designing a new gene therapy strategy is to identify the gene or protein of therapeutic value for a given disease. In monogenic disorders like cystic fibrosis (CF), the answer is obvious, but for a multifactorial disease like cancer, this is rarely the case. The identification of a particular cell type or cell types as the target population for gene therapy must also be identified early in the design phase. This issue implies knowledge of where the protein of interest must function for therapeutic effect. It also implies an understanding of how the protein may or may not be distributed to its site of action from any given site of production.

As one example of this consideration, CF is a disease of the conducting airways in which the transgene product is an integral membrane protein, which is unable to spread to other compartments once it is expressed within the target cells (Goldman et al., 1997; Limberis et al., 2002; Lim et al., 2003; Sinn et al., 2003). Thus, the airway epithelium itself must be targeted for gene transfer. This is certainly not the case in alpha-1 antitrypsin (AAT) deficiency (A1AD), however. In this disease, injury to cells and extracellular matrix occur within the alveolar interstitium and perhaps also in the airways. Because the AAT protein is secreted into the extracellular space, gene transfer can potentially be targeted to a number of different cell types and still retain its ability to generate a therapeutic effect. These potential targets may be either within the lung itself or at some distant body site.

The potential success of any gene therapy strategy is, therefore, dependent upon the state of scientific knowledge as to the necessary site of action of the transgene product and on its ability to be transported between compartments. Since the permissiveness to rAAV-mediated gene transfer varies widely between different cell types within different compartments, this initial analysis can greatly affect the potential for success. In the recent years, with the availability of new rAAV serotypes and capsid mutants, this might also dictate the choice of the variety of rAAV vector used.

For the purposes of this discussion, the lung may be viewed as consisting of the following compartments:
- Alveoli
- Airways
- Other: Vascular structures and pleura

Although this classification is somewhat arbitrary and there are numerous individual cell types within each of these compartments, this consideration is meant to be functional with regard to potential gene transfer targets, rather than comprehensive in the context of pulmonary anatomy. In that context, one begins to consider individual disease states in terms of which compartment or compartments must serve as targets for gene transfer.

4.3. Therapies targeting the alveoli

Alveoli represent the primary site for gas exchange within the lung, and thus their health is vital for survival. Alveolar conditions with a primary genetic cause, such as surfactant protein-B (SP-B) deficiency and SP-C deficiency, are prime candidates for a rAAV-gene therapy approach. Diseases in which alveoli are damaged secondary to other defects might also be treated with gene transfer. Such conditions include environmental toxin exposure, infectious diseases, and adult respiratory distress syndrome (ARDS) (Table 4.1) (Kolls et al., 1997, 1998, 2001; Cheers et al., 1999; Ruan et al., 2002).

Early reports with rAd vectors may provide the proof of principle for transient expression of a number of potentially therapeutic proteins. For instance the Na^+/K^+-ATPase was found to promote reabsorption of alveolar fluid in an ARDS model (Factor et al., 1999; Sartori and Matthay, 2002). Anti-inflammatory cytokines and antioxidants have also been used with some success in a similar context in animal models of sepsis and ARDS (Van Laethem et al., 1998; Dumasius et al., 2002).

TABLE 4.1
Gene therapy targets from the alveolar compartment

Genetic	Inflammatory	Infectious
Alpha-1 antitrypsin deficiency	Adult respiratory distress syndrome (ARDS)	Pneumocystis pneumonia
Surfactant protein deficiency (SP-B, SP-C)	Idiopathic interstitial pulmonary fibrosis	Bacterial pneumonia
Alveolar proteinosis	Interstitial pneumonitis	Tuberculosis

One instance of a disease affecting the alveolar interstitium is A1AD. This common monogenic disorder is caused by a lack of functional AAT, the major circulating serum antiprotease that normally protects the delicate elastin fibers of the alveolar interstitium from destruction by neutrophil elastase. AAT also has activity to counteract other neutrophil by-products, including cathespin-G, proteinase-3, and alpha defensins. The multiple protective functions of AAT in the lung can be provided by replacing serum levels of AAT up to the range of 11 micromolar (roughly 570–800 mg/ml). These values were derived from genotype–phenotype data (Crystal, 1990).

There are two protein replacement products available for A1AD, both pooled human plasma donor-derived products. Approval of these products was based upon replacement of serum AAT to the above-mentioned levels. Neither has been proven to prevent lung disease in a prospective, double-blinded placebo-controlled fashion, but retrospective registry data suggest a beneficial effect. However, protein replacement therapy has been very safe. Particularly remarkable has been the lack of an adaptive immune response by patients on therapy, which may relate to the genetic homogeneity of this A1AD population, in which approximately 95% have one particular missense mutant allele (PI*Z).

The rationale for gene therapy of A1AD is strongly supported by the safety and efficacy of the protein replacement. rAAV-produced AAT is efficiently secreted from a number of different sites in mice, including the liver (which is its natural site of production), the skeletal muscle, the airway, the pleura, and the peritoneum. Any of these sites could potentially be chosen for a gene augmentation strategy for A1AD lung disease. In the minority of patients with clinically evident A1AD liver disease this kind of approach may not be valid. This subgroup of individuals is thought to undergo hepatocellular damage due to misfolding and/or polymerization of the mutant protein. Therefore, gene therapy with normal (PI*M) AAT may not be therapeutic for the liver disease unless gene transfer were accompanied by repression of mutant protein expression.

Treatment of A1AD lung disease is a much simpler task, involving simple sustained augmentation of serum levels. In moving toward that goal, pre-clinical work supports the concept of intramuscular administration of rAAV2-AAT. Reports from mouse, rabbit, and baboon models all indicate safety and bioactivity of this vector (Song et al., 1998, 2001, 2002; Poirier et al., 2004). Based upon that data, a phase I clinical trial in AAT deficient adults was recently initiated (Flotte et al., 2004).

Theoretically, a number of other diseases affecting the pulmonary interstitium might be treated with gene therapy. These include interstitial pneumonitides and idiopathic pulmonary fibrosis. These approaches will depend upon further advancements in the study of the basic pathobiology of these disorders.

4.4. Therapies targeting the airways

Table 4.2 lists examples of diseases of the conducting airways that might be approached via gene therapy. The list includes purely genetic disorders like primary ciliary dyskinesias (PCD) and CF, multifactorial diseases, like asthma, and a variety of infectious diseases, such as influenza and respiratory syncytial virus. The conducting airways consist of the nose, pharynx, larynx, trachea, bronchi (approximately eight generations of branching), and bronchioles

TABLE 4.2
Gene therapy targets from the airway compartment

Genetic	Inflammatory	Infectious	Neoplastic
Cystic fibrosis	Asthma	Respiratory syncytial virus	Carcinoma
Primary ciliary dyskinesias		Influenza	

(another approximately eight generations of asymmetric dichotomous branching, ending with the terminal bronchioles). Bronchi are distinguished from bronchioles by the presence of at least partial cartilagenous support and submucosal glands. The terminal bronchioles are the most distal units within the lung that are surrounded by smooth muscle. In the more proximal conducting airways the epithelium is organized as a pseudostratified columnar epithelium, with the exception of the pharynx, which is lined with squamous epithelium. As one progresses more distally through the trachea and bronchi to the bronchioles, the high columnar epithelium becomes progressively more cuboidal, down to a true simple ciliated cuboidal epithelium in the terminal bronchioles. Interspersed within the epithelium between the ciliated cells are a variety of secretory cells, including the mucous or goblet cells, serous secretory cells, basal cells, neuroendocrine cells, and (in the more distal airways) Clara cells.

In CF, there is consensus in favor of targeting the surface ciliated epithelium as the primary target for gene transfer, but there is also clear evidence that the serous cells of the submucosal glands represent "hot spots" for native CFTR expression. It is not clear whether or not that means that the submucosal gland expression of CFTR will be required for clinical efficacy. There has also been a great deal of discussion about the presence of progenitor cells within the airways. In general, there has been a prevailing opinion that basal cells in the more proximal airways and Clara cells in the peripheral airways may serve as progenitors. In addition, there is evidence that ciliated columnar cells can transdifferentiate into secretory cells and vice versa. The potential for a cell type to serve as a progenitor might have importance for its ability to serve as a target for long-term gene transfer, if a strategy to enhance vector integration can be devised.

4.4.1. Clinical gene therapy for cystic fibrosis (CF)

In the mid-1990s a number of gene therapy trials were initiated in CF patients, including the first human trials of recombinant adeno-associated virus serotype 2 (rAAV2). The rAd-CFTR trials had been initiated two years earlier, and quickly generated data demonstrating the feasibility of gene expression (Rosenfeld et al., 1992). However, gene transfer was short-term and was associated with an immediate cytokine-mediated inflammatory response, which was apparently dose-limiting (Brody et al., 1994; Crystal et al., 1994; Ben-Gary et al., 2002). Preliminary studies suggested that correction of the transepithelial potential difference abnormalities in CF might be feasible with rAd vectors (Zabner et al., 1993), but later studies with a control rAd vector indicated that vector-mediated cytotoxicity may have confounded those results (Knowles et al., 1995) by leading to general leakiness of the nasal epithelium. Trials of cationic liposomes similarly demonstrated that transient gene transfer was feasible in the nose (Alton et al., 1993; Caplen et al., 1995; Porteous et al., 1997; Knowles et al., 1998; Hyde et al., 2000). Another study demonstrated that naked DNA itself was capable of gene transfer (Zabner et al., 1997) in the nasal epithelium. Aerosol inhalation of cationic liposome-DNA complexes was limited, however, by adverse reactions, including a flu-like syndrome elicited after aerosol inhalation of lipoplexes in CF patients (McLachlan et al., 2000; Ruiz et al., 2001).

Recombinant AAV serotype 2 vectors were developed for CFTR gene transfer. These vectors were designed to overcome the intrinsic packaging limitation of rAAV by fitting the relatively large coding sequence of CFTR (approximately 4.5 kb), with minimal promoter sequences intrinsic to the AAV genome. A series of pre-clinical (Flotte et al., 1993a,b; Afione et al., 1996; Conrad et al., 1996; Beck et al., 2002) and phase I clinical trials were performed, which demonstrated safety in the nose (Flotte et al., 2003), bronchus (Flotte et al., 2003), paranasal sinus (Wagner et al., 1998, 1999), and after inhalation as a nebulized aerosol (Aitken et al., 2001). Dose-related gene transfer was also observed. Early phase II clinical trials of this rAAV2-CFTR vector were performed in the sinus and the lower respiratory tract. Both studies indicated a decline in the proinflammatory cytokine, IL-8. There was improvement in CFTR function as demonstrated by electrophysiology in the sinuses, and an improvement in airway function as demonstrated by an increase in FEV-1 in the phase II aerosol trial (Moss et al., 2004). Unfortunately, the positive effects did not last longer than 30 days.

4.4.2. Barriers to rAAV2 vectors in the airways

Apart from the promise of rAAV2-CFTR gene transfer, the effort to make AAV2 into a vector for CF also led to a better understanding of a number of key limitations to rAAV2 in the airways. Among these were extracellular barriers, such as neutrophil elastase, neutrophil alpha defensins, and neutralizing antibodies. At the cell surface other barriers were encountered. The AAV2 receptors (heparan sulfate proteoglycan, fibroblast growth factor receptor, and alpha$_v$beta$_5$ integrins) were found to be scarce at the apical surface of the airway cells. After entry past the cell surface, the vector encountered other barriers, including degradation of vector particles in the proteosome and inhibition of conversion of vector DNA to its active double-stranded form by FKBP52, a host DNA-binding protein. Finally, rAAV also has an intrinsic limitation due to the small net packaging capacity of this virion (4.7 kb) relative to the size of the CFTR coding sequence (4.43 kb) (Flotte et al., 1993b; Duan et al., 2000, 2001).

TABLE 4.3
Limitations and solutions for rAAV2 in the airways

Limitation	Strategy to circumvent
Extracellular barriers (DNA, enzymes, mucus)	Airway clearance, anti-inflammatory, anti-protease pre-treatments
Paucity of receptors	Alternate serotypes, targeted capsid mutants
Proteosome-mediated degradation	Proteosome inhibitors (tripeptides, anthracyclines)
Inhibition of second-strand synthesis	Tyrosine kinase inhibitors

The greatly increased knowledge of vector limitations relative to airway gene therapy has led to a number of key innovations in vector design (Table 4.3). These include the introduction of new rAAV serotypes and capsid mutants capable of targeting other receptors that may be present in greater abundance on the apical surface of airway cells (Bartlett et al., 1999; Wu et al., 2000; Zabner et al., 2000; Gao et al., 2002, 2003; Loiler et al., 2003). Post-entry blockades may be amenable to intervention as well. For instance, proteosome inhibitors have been shown to greatly accentuate the efficacy of rAAV-CFTR gene transfer in a number of in vitro models (Yan et al., 2002). Some of these agents are already approved for use in humans in other conditions, and so could be incorporated fairly readily into future gene therapy trials. Finally, there is also indication that tyrosine kinase inhibitors could maximize the transcriptional activity of rAAV vectors by inhibiting the phosphorylation of FKBP52.

4.4.3. rAAV gene transfer to the airways as therapy for other diseases

As with other organ systems, the consideration of using gene therapy in the airways has generally begun with simple Mendelian disorders, like CF. However, asthma is a much more prevalent multifactorial

TABLE 4.4
Potential targets for asthma gene therapy

Th-1 biasing	Inhibitors of effector response
Interferon-gamma	IL-13 inhibitor (soluble receptor-fusion)
Interleukin-2	Dominant negative mutant IL-4
Interleukin-12	IL-10
CpG-oligonucleotides	Galectin-3
Interleukin-18	

condition that has also been investigated in pre-clinical models. A number of reports have shown that Th1-related cytokines will lessen the severity of inflammation in an ovalbumin-sensitized mouse model of asthma (Kumar et al., 1999, 2003; Walter et al., 2001; Kline, 2002; Nishikubo et al., 2003; Zavorotinskaya et al., 2003). Another approach would be to block the effector cells of the asthmatic response (del Pozo et al., 2002), many of which are derived from Th2 lymphocytes. An ever-expanding list of gene and gene products has been considered as possible therapies for asthma (Table 4.4).

4.5. Therapies targeting the pulmonary vasculature and pleura

Pulmonary hypertension is a devastating, potentially fatal disorder. Recent years have witnessed a great expansion in our understanding of the molecular pathophysiology of this condition. Options for therapy have just become available in recent years, focused either on the prostacyclin pathway or the nitric oxide pathway for pulmonary arteriolar relaxation. Continuous infusion of prostacyclin has been successful, demonstrating the relevance of the former pathway. The efficacy of both inhaled nitric oxide and systemic sildenafil support the latter. In the future, gene transfer could be used for sustained delivery of either of these agents, by means of augmenting the expression of prostacyclin synthase or nitric oxide synthase. Recent discoveries demonstrating a connection between the bone morphogenic protein receptor type 2 (BMP-R-II) gene and familial primary pulmonary hypertension has raised the possibility of targeting BMP-R-II in such cases (Eddahibi et al., 2002; Morse, 2002).

The pleura also represents a critical compartment for gene transfer therapies. In fact, several cancer gene therapy approaches have targeted pleura for treatment of mesothelioma, a malignancy with an overall poor prognosis (Elshami et al., 1996; Schwarzenberger et al., 1998a,b; Sterman et al., 1998; Albelda et al., 2002; Friedlander et al., 2003). A number of key phase I trials and some later phase trials have been undertaken in patients with mesothelioma using rAd vectors. It remains to be seen whether rAAV vectors might be useful in this context.

4.6. Future directions

The potential for CF gene therapy with rAAV vectors remains a contradiction. While multiple new efforts are underway to promote clinical development of new serotypes and augmentation strategies, such as proteosome inhibitors, this disease target remains very difficult for a number of reasons. First, there is an intrinsic difficulty with delivering a gene therapy for a protein that must function as an intrinsic membrane protein within the epithelium of most or all of the lung. The very widespread efficiency and uniform distribution that might be required is rendered all the more problematic by the fact that these patients develop lung disease very early in life. It appears that CF lung disease hinders uniform distribution of therapeutic agents and increases the barriers to gene transfer in terms of extracellular secretions. Ultimately, the field may be forced to consider a gene transfer strategy targeted early in life. The episomal nature of current rAAV vectors presents another problem for CF gene therapy. Even if one were to get an ideal distribution and gene transfer to all regions of the lung, unless one would be able to achieve vector integration into progenitor cells capable of repopulating the airways, the gene therapy is very likely to be transient. There is hope for the development of new AAV vectors that facilitate Rep-mediated, site-specific integration, but these seem to be several years away from their clinical application. The near-term outlook for successful gene therapy for deficiencies of secreted proteins, like A1AD is brighter. In the absence of injury, muscle fibers and hepatocytes remain in G_0 and the rAAV vector genomes appear to remain and retain their ability to express the transgene. If this is proven to occur in humans, it has the potential to provide a paradigm for simple, yet effective gene therapy strategies.

Acknowledgements

Supported in part by grants from the NCRR (RR00082, RR16586), the NHLBI (HL69877, HL51811, HL59412), the NIDDK (DK583237), the NEI (EY13729), the Alpha One Foundation, and the Cystic Fibrosis Foundation.

References

Afione, S. A., Conrad, C. K., Kearns, W. G., Chunduru, S., Adams, R., Reynolds, T. C., Guggino, W. B., Cutting, G. R., Carter, B. J. and Flotte, T. R. (1996). In vivo model of adeno-associated virus vector persistence and rescue. J. Virol. *70*, 3235–3241.

Aitken, M. L., Moss, R. B., Waltz, D. A., Dovey, M. E., Tonelli, M. R., McNamara, S. C., Gibson, R. L., Ramsey, B. W., Carter, B. J. and Reynolds, T. C. (2001). A phase I study of aerosolized administration of tgAAVCF to cystic fibrosis subjects with mild lung disease. Hum. Gene Ther. *12*, 1907–1916.

Albelda, S. M., Wiewrodt, R. and Sterman, D. H. (2002). Gene therapy for lung neoplasms. Clin. Chest Med. *23*, 265–277.

Alton, E. W., Middleton, P. G., Caplen, N. J., Smith, S. N., Steel, D. M., Munkonge, F. M., Jeffery, P. K., Geddes, D. M., Hart, S. L. Williamson, R.

et al. (1993). Non-invasive liposome-mediated gene delivery can correct the ion transport defect in cystic fibrosis mutant mice [published erratum appears in Nat. Genet. 1993 Nov. *5(3)*, 312]. Nat. Genet. *5*, 135–142.

Beck, S. E., Laube, B. L., Barberena, C. I., Fischer, A. C., Adams, R. J., Chesnut, K., Flotte, T. R. and Guggino, W. B. (2002). Deposition and expression of aerosolized rAAV vectors in the lungs of Rhesus macaques. Mol. Ther. *6*, 546–554.

Bartlett, J. S., Kleinschmidt, J., Boucher, R. C. and Samulski, R. J. (1999). Targeted adeno-associated virus vector transduction of nonpermissive cells mediated by a bispecific F(ab'gamma)2 antibody. Nat. Biotechnol. *17*, 181–186.

Ben-Gary, H., McKinney, R. L., Rosengart, T., Lesser, M. L. and Crystal, R. G. (2002). Systemic interleukin-6 responses following administration of adenovirus gene transfer vectors to humans by different routes. Mol. Ther. *6*, 287–297.

Brody, S. L., Metzger, M., Danel, C., Rosenfeld, M. A. and Crystal, R. G. (1994). Acute responses of non-human primates to airway delivery of an adenovirus vector containing the human cystic fibrosis transmembrane conductance regulator cDNA. Hum. Gene Ther. *5*, 821–836.

Crystal, R. G. (1990). Alpha 1-antitrypsin deficiency, emphysema, and liver disease Genetic basis and strategies for therapy. J. Clin. Invest. *85*, 1343–1352.

Crystal, R. G., McElvaney, N. G., Rosenfeld, M. A., Chu, C. S., Mastrangeli, A., Hay, J. G., Brody, S. L., Jaffe, H. A., Eissa, N. T. and Danel, C. (1994). Administration of an adenovirus containing the human CFTR cDNA to the respiratory tract of individuals with cystic fibrosis [see comments]. Nat. Genet. *8*, 42–51.

Caplen, N. J., Alton, E. W., Middleton, P. G., Dorin, J. R., Stevenson, B. J., Gao, X., Durham, S. R., Jeffery, P. K., Hodson, M. E. Coutelle, C. et al. (1995). Liposome-mediated CFTR gene transfer to the nasal epithelium of patients with cystic fibrosis [see comments] [published erratum appears in Nat. Med. 1995 Mar. *1(3)*, 272]. Nat. Med. *1*, 39–46.

Cheers, C., Janas, M., Ramsay, A. and Ramshaw, I. (1999). Use of recombinant viruses to deliver cytokines influencing the course of experimental bacterial infection. Immunol. Cell. Biol. *77*, 324–330.

Conrad, C. K., Allen, S. S., Afione, S. A., Reynolds, T. C., Beck, S. E., Fee-Maki, M., Barrazza-Ortiz, X., Adams, R., Askin, F. B., Carter, B. J., Guggino, W. B. and Flotte, T. R. (1996). Safety of single-dose administration of an adeno-associated virus (AAV)-CFTR vector in the primate lung. Gene Ther. *3*, 658–668.

del Pozo, V., Rojo, M., Rubio, M. L., Cortegano, I., Cardaba, B., Gallardo, S., Ortega, M., Civantos, E., Lopez, E., Martin-Mosquero, C., Peces-Barba, G., Palomino, P., Gonzalez-Mangado, N. and Lahoz, C. (2002). Gene therapy with galectin-3 inhibits bronchial obstruction and inflammation in antigen-challenged rats through interleukin-5 gene downregulation. Am. J. Respir. Crit. Care Med. *166*, 732–737.

Duan, D., Yue, Y. and Engelhardt, J. F. (2001). Expanding AAV packaging capacity with trans-splicing or overlapping vectors: A quantitative comparison. Mol. Ther. *4*, 383–391.

Duan, D., Yue, Y., Yan, Z. and Engelhardt, J. F. (2000). A new dual-vector approach to enhance recombinant adeno-associated virus-mediated gene expression through intermolecular cis activation. Nat. Med. *6*, 595–598.

Dumasius, V., Mendez, M., Mutlu, G. M. and Factor, P. (2002). Acute lung injury does not impair adenoviral-mediated gene transfer to the alveolar epithelium. Chest *121*, 33S–34S.

Eddahibi, S., Morrell, N., d'Ortho, M. P., Naeije, R. and Adnot, S. (2002). Pathobiology of pulmonary arterial hypertension. Eur. Respir. J. *20*, 1559–1572.

Elshami, A. A., Kucharczuk, J. C., Zhang, H. B., Smythe, W. R., Hwang, H. C., Litzky, L. A., Kaiser, L. R. and Albelda, S. M. (1996). Treatment of pleural mesothelioma in an immunocompetent rat model utilizing adenoviral transfer of the herpes simplex virus thymidine kinase gene. Hum. Gene Ther. *7*, 141–148.

Factor, P., Dumasius, V., Saldias, F. and Sznajder, J. I. (1999). Adenoviral-mediated overexpression of the NA,K-ATPase beta1 subunit gene increases lung edema clearance and improves survival during acute hyperoxic lung injury in rats. Chest *116*, 24S–25S.

Flotte, T. R., Afione, S. A., Conrad, C., McGrath, S. A., Solow, R., Oka, H., Zeitlin, P. L., Guggino, W. B. and Carter, B. J. (1993a). Stable in vivo expression of the cystic fibrosis transmembrane conductance regulator with an adeno-associated virus vector. Proc. Natl. Acad. Sci. USA *90*, 10613–10617.

Flotte, T. R., Afione, S. A., Solow, R., Drumm, M. L., Markakis, D., Guggino, W. B., Zeitlin, P. L. and Carter, B. J. (1993b). Expression of the cystic fibrosis transmembrane conductance regulator from a novel adeno-associated virus promoter. J. Biol. Chem. *268*, 3781–3790.

Flotte, T. R., Brantly, M. L., Spencer, L. T., Byrne, B. J., Spencer, C. T., Baker, D. J. and Humphries, M. (2004). Phase I trial of intramuscular injection of a recombinant adeno-associated virus alpha 1-antitrypsin (rAAV2-CB-hAAT) gene vector to AAT-deficient adults. Hum. Gene Ther. *15*, 93–128.

Flotte, T. R., Carter, B., Conrad, C., Guggino, W., Reynolds, T., Rosenstein, B., Taylor, G., Walden, S. and Wetzel, R. (1996). A phase I study of an adeno-associated virus-CFTR gene vector in adult CF patients with mild lung disease. Hum. Gene Ther. *7*, 1145–1159.

Flotte, T. R., Zeitlin, P. L., Reynolds, T. C., Heald, A. E., Pedersen, P., Beck, S., Conrad, C. K., Brass-Ernst, L., Humphries, M., Sullivan, K., Wetzel, R., Taylor, G., Carter, B. J. and Guggino, W. B. (2003). Phase I trial of intranasal and endobronchial administration of a recombinant adeno-associated virus serotype 2 (rAAV2)-CFTR vector in adult cystic fibrosis patients: A two-part clinical study. Hum. Gene. Ther. *14*, 1079–1088.

Friedlander, P. L., Delaune, C. L., Abadie, J. M., Toups, M., La Cour, J., Marrero, L., Zhong, Q. and Kolls, J. K. (2003). Efficacy of CD40 ligand gene therapy in malignant mesothelioma. Am. J. Respir. Cell. Mol. Biol. *29*, 321–330.

Gao, G., Alvira, M. R., Somanathan, S., Lu, Y., Vandenberghe, L. H., Rux, J. J., Calcedo, R., Sanmiguel, J., Abbas, Z. and Wilson, J. M. (2003). Adeno-associated viruses undergo substantial evolution in primates during natural infections. Proc. Natl. Acad. Sci. USA *100*, 6081–6086.

Gao, G. P., Alvira, M. R., Wang, L., Calcedo, R., Johnston, J. and Wilson, J. M. (2002). Novel adeno-associated viruses from rhesus monkeys as vectors for human gene therapy. Proc. Natl. Acad. Sci. USA *99*, 11854–11859.

Goldman, M. J., Lee, P. S., Yang, J. S. and Wilson, J. M. (1997). Lentiviral vectors for gene therapy of cystic fibrosis. Hum. Gene Ther. *8*, 2261–2268.

Hyde, S. C., Southern, K. W., Gileadi, U., Fitzjohn, E. M., Mofford, K. A., Waddell, B. E., Gooi, H. C., Goddard, C. A., Hannavy, K., Smyth, S. E., Egan, J. J., Sorgi, F. L., Huang, L., Cuthbert, A. W., Evans, M. J., Colledge, W. H., Higgins, C. F., Webb, A. K. and Gill, D. R. (2000). Repeat administration of DNA/liposomes to the nasal epithelium of patients with cystic fibrosis. Gene Ther. *7*, 1156–1165.

Kline, J. N. (2002). DNA therapy for asthma. Curr. Opin. Allergy Clin. Immunol. *2*, 69–73.

Knowles, M. R., Hohneker, K. W., Zhou, Z., Olsen, J. C., Noah, T. L., Hu, P. C., Leigh, M. W., Engelhardt, J. F., Edwards, L. J. Jones, K. R. et al. (1995). A controlled study of adenoviral-vector-mediated gene transfer in the nasal epithelium of patients with cystic fibrosis [see comments]. N. Engl. J. Med. *333*, 823–831.

Knowles, M. R., Noone, P. G., Hohneker, K., Johnson, L. G., Boucher, R. C., Efthimiou, J., Crawford, C., Brown, R., Schwartzbach, C. and Pearlman, R. (1998). A double-blind, placebo controlled, dose ranging study to evaluate the safety and biological efficacy of the lipid-DNA complex

GR213487B in the nasal epithelium of adult patients with cystic fibrosis. Hum. Gene Ther. *9*, 249–269.
Kolls, J. K., Lei, D., Nelson, S., Summer, W. R. and Shellito, J. E. (1997). Pulmonary cytokine gene therapy. Adenoviral-mediated murine interferon gene transfer compartmentally activates alveolar macrophages and enhances bacterial clearance. Chest *111*, 104S.
Kolls, J. K., Lei, D., Stoltz, D., Zhang, P., Schwarzenberger, P. O., Ye, P., Bagby, G., Summer, W. R., Shellito, J. E. and Nelson, S. (1998). Adenoviral-mediated interferon-gamma gene therapy augments pulmonary host defense of ethanol-treated rats. Alcohol Clin. Exp. Res. *22*, 157–162.
Kolls, J. K., Ye, P. and Shellito, J. E. (2001). Gene therapy to modify pulmonary host defenses. Semin. Respir. Infect. *16*, 18–26.
Kumar, M., Behera, A. K., Matsuse, H., Lockey, R. F. and Mohapatra, S. S. (1999). Intranasal IFN-gamma gene transfer protects BALB/c mice against respiratory syncytial virus infection. Vaccine *18*, 558–567.
Kumar, M., Kong, X., Behera, A. K., Hellermann, G. R., Lockey, R. F. and Mohapatra, S. S. (2003). Chitosan IFN-gamma-pDNA nanoparticle (CIN) therapy for allergic asthma. Genet. Vaccines Ther. *1*, 3.
Lim, F. Y., Kobinger, G. P., Weiner, D. J., Radu, A., Wilson, J. M. and Crombleholme, T. M. (2003). Human fetal trachea-SCID mouse xenografts: Efficacy of vesicular stomatitis virus-G pseudotyped lentiviral-mediated gene transfer. J. Pediatr. Surg. *38*, 834–839.
Limberis, M., Anson, D. S., Fuller, M. and Parsons, D. W. (2002). Recovery of airway cystic fibrosis transmembrane conductance regulator function in mice with cystic fibrosis after single-dose lentivirus-mediated gene transfer. Hum. Gene Ther. *13*, 1961–1970.
Loiler, S. A., Conlon, T. J., Song, S., Tang, Q., Warrington, K. H., Agarwal, A., Kapturczak, M., Li, C., Ricordi, C., Atkinson, M. A., Muzyczka, N. and Flotte, T. R. (2003). Targeting recombinant adeno-associated virus vectors to enhance gene transfer to pancreatic islets and liver. Gene Ther. *10*, 1551–1558.
McLachlan, G., Stevenson, B. J., Davidson, D. J. and Porteous, D. J. (2000). Bacterial DNA is implicated in the inflammatory response to delivery of DNA/DOTAP to mouse lungs. Gene Ther. *7*, 384–392.
Morse, J. H. (2002). Bone morphogenetic protein receptor 2 mutations in pulmonary hypertension. Chest *121*, 50S–53S.
Moss, R. B., Rodman, D., Spencer, L. T., Aitken, M. L., Zeitlin, P. L., Waltz, D., Milla, C., Brody, A. S., Clancy, J. P., Ramsey, B., Hamblett, N. and Heald, A. E. (2004). Repeated adeno-associated virus serotype 2 aerosol-mediated cystic fibrosis transmembrane regulator gene transfer to the lungs

of patients with cystic fibrosis: A multicenter, double-blind, placebo-controlled trial. Chest 125, 509–521.
Nishikubo, K., Murata, Y., Tamaki, S., Sugama, K., Imanaka-Yoshida, K., Yuda, N., Kai, M., Takamura, S., Sebald, W., Adachi, Y. and Yasutomi, Y. (2003). A single administration of interleukin-4 antagonistic mutant DNA inhibits allergic airway inflammation in a mouse model of asthma. Gene Ther. 10, 2119–2125.
Poirier, A., Campbell-Thompson, M., Tang, Q., Scott-Jorgensen, M., Combee, L., Loiler, S., Crawford, J., Song, S. and Flotte, T. R. (2004). Toxicology and biodistribution studies of a recombinant adeno-associated virus 2-alpha-1 antitrypsin vector. Preclinica 2, 43–51.
Porteous, D. J., Dorin, J. R., McLachlan, G., Davidson-Smith, H., Davidson, H., Stevenson, B. J., Carothers, A. D., Wallace, W. A., Moralee, S., Hoenes, C., Kallmeyer, G., Michaelis, U., Naujoks, K., Ho, L. P., Samways, J. M., Imrie, M., Greening, A. P. and Innes, J. A. (1997). Evidence for safety and efficacy of DOTAP cationic liposome mediated CFTR gene transfer to the nasal epithelium of patients with cystic fibrosis. Gene Ther. 4, 210–218.
Rosenfeld, M. A., Yoshimura, K., Trapnell, B. C., Yoneyama, K., Rosenthal, E. R., Dalemans, W., Fukayama, M., Bargon, J., Stier, L. E. Stratford-Perricaudet, L. et al. (1992). In vivo transfer of the human cystic fibrosis transmembrane conductance regulator gene to the airway epithelium. Cell 68, 143–155.
Ruan, S., Tate, C., Lee, J. J., Ritter, T., Kolls, J. K. and Shellito, J. E. (2002). Local delivery of the viral interleukin-10 gene suppresses tissue inflammation in murine Pneumocystis carinii infection. Infect. Immun. 70, 6107–6113.
Ruiz, F. E., Clancy, J. P., Perricone, M. A., Bebok, Z., Hong, J. S., Cheng, S. H., Meeker, D. P., Young, K. R., Schoumacher, R. A., Weatherly, M. R., Wing, L., Morris, J. E., Sindel, L., Rosenberg, M., van Ginkel, F. W., McGhee, J. R., Kelly, D., Lyrene, R. K. and Sorscher, E. J. (2001). A clinical inflammatory syndrome attributable to aerosolized lipid-DNA administration in cystic fibrosis. Hum. Gene Ther. 12, 751–761.
Sartori, C. and Matthay, M. A. (2002). Alveolar epithelial fluid transport in acute lung injury: New insights. Eur. Respir. J. 20, 1299–1313.
Schwarzenberger, P., Harrison, L., Weinacker, A., Marrogi, A., Byrne, P., Ramesh, R., Theodossiou, C., Gaumer, R., Summer, W., Freeman, S. M. and Kolls, J. K. (1998). The treatment of malignant mesothelioma with a gene modified cancer cell line: A phase I study. Hum. Gene. Ther. 9, 2641–2649.

Schwarzenberger, P., Harrison, L., Weinacker, A., Gaumer, R., Theodossiou, C., Summer, W., Ye, P., Marrogi, A. J., Ramesh, R., Freeman, S. and Kolls, J. (1998). Gene therapy for malignant mesothelioma: A novel approach for an incurable cancer with increased incidence in Louisiana. J. La State Med. Soc. *150*, 168–174.

Sinn, P. L., Hickey, M. A., Staber, P. D., Dylla, D. E., Jeffers, S. A., Davidson, B. L., Sanders, D. A. and McCray, P. B., Jr. (2003). Lentivirus vectors pseudotyped with filoviral envelope glycoproteins transduce airway epithelia from the apical surface independently of folate receptor alpha. J. Virol. *77*, 5902–5910.

Song, S., Morgan, M., Ellis, T., Poirier, A., Chesnut, K., Wang, J., Brantly, M., Muzyczka, N., Byrne, B. J., Atkinson, M. and Flotte, T. R. (1998). Sustained secretion of human alpha–1-antitrypsin from murine muscle transduced with adeno-associated virus vectors. Proc. Natl. Acad. Sci. USA *95*, 14384–14388.

Song, S., Laipis, P. J., Berns, K. I. and Flotte, T. R. (2001). Effect of DNA-dependent protein kinase on the molecular fate of the rAAV2 genome in skeletal muscle. Proc. Natl. Acad. Sci. USA *98*, 4084–4088.

Song, S., Scott-Jorgensen, M., Wang, J., Poirier, A., Crawford, J., Campbell-Thompson, M. and Flotte, T. R. (2002). Intramuscular administration of recombinant adeno-associated virus 2 alpha–1 antitrypsin (rAAV-SERPINA1) vectors in a nonhuman primate model: Safety and immunologic aspects. Mol. Ther. *6*, 329–335.

Sterman, D. H., Treat, J., Litzky, L. A., Amin, K. M., Coonrod, L., Molnar-Kimber, K., Recio, A., Knox, L., Wilson, J. M., Albelda, S. M. and Kaiser, L. R. (1998). Adenovirus-mediated herpes simplex virus thymidine kinase/ ganciclovir gene therapy in patients with localized malignancy: Results of a phase I clinical trial in malignant mesothelioma. Hum. Gene Ther. *9*, 1083–1092.

Van Laethem, J. L., Eskinazi, R., Louis, H., Rickaert, F., Robberecht, P. and Deviere, J. (1998). Multisystemic production of interleukin 10 limits the severity of acute pancreatitis in mice. Gut *43*, 408–413.

Wagner, J. A., Messner, A. H., Moran, M. L., Daifuku, R., Kouyama, K., Desch, J. K., Manly, S., Norbash, A. M., Conrad, C. K., Friborg, S., Reynolds, T., Guggino, W. B., Moss, R. B., Carter, B. J., Wine, J. J., Flotte, T. R. and Gardner, P. (1999). Safety and biological efficacy of an adeno-associated virus vector-cystic fibrosis transmembrane conductance regulator (AAV-CFTR) in the cystic fibrosis maxillary sinus. Laryngoscope *109*, 266–274.

Wagner, J. A., Reynolds, T., Moran, M. L., Moss, R. B., Wine, J. J., Flotte, T. R. and Gardner, P. (1998). Efficient and persistent gene transfer of AAV-CFTR in maxillary sinus. Lancet *351*, 1702–1703.

Walter, D. M., Wong, C. P., De Kruyff, R. H., Berry, G. J., Levy, S. and Umetsu, D. T. (2001). Il-18 gene transfer by adenovirus prevents the development of and reverses established allergen-induced airway hyperreactivity. J. Immunol. *166*, 6392–6398.

Wang, G., Sinn, P. L. and McCray, P. B., Jr. (2000). Development of retroviral vectors for gene transfer to airway epithelia. Curr. Opin. Mol. Ther. *2*, 497–506.

Wu, P., Xiao, W., Conlon, T., Hughes, J., Agbandje-McKenna, M., Ferkol, T., Flotte, T. R. and Muzyczka, N. (2000). Mutational analysis of the adeno-associated virus type 2 (AAV2) capsid gene and construction of AAV2 vectors with altered tropism [In Process Citation]. J. Virol. *74*, 8635–8647.

Yan, Z., Zak, R., Luxton, G. W., Ritchie, T. C., Bantel-Schaal, U. and Engelhardt, J. F. (2002). Ubiquitination of both adeno-associated virus type 2 and 5 capsid proteins affects the transduction efficiency of recombinant vectors. J. Virol. *76*, 2043–2053.

Zabner, J., Cheng, S. H., Meeker, D., Launspach, J., Balfour, R., Perricone, M. A., Morris, J. E., Marshall, J., Fasbender, A., Smith, A. E. and Welsh, M. J. (1997). Comparison of DNA-lipid complexes and DNA alone for gene transfer to cystic fibrosis airway epithelia in vivo. J. Clin. Invest. *100*, 1529–1537.

Zabner, J., Couture, L. A., Gregory, R. J., Graham, S. M., Smith, A. E. and Welsh, M. J. (1993). Adenovirus-mediated gene transfer transiently corrects the chloride transport defect in nasal epithelia of patients with cystic fibrosis. Cell *75*, 207–216.

Zabner, J., Seiler, M., Walters, R., Kotin, R. M., Fulgeras, W., Davidson, B. L. and Chiorini, J. A. (2000). Adeno-associated virus type 5 (AAV5) but not AAV2 binds to the apical surfaces of airway epithelia and facilitates gene transfer. J. Virol. *74*, 3852–3858.

Zavorotinskaya, T., Tomkinson, A. and Murphy, J. E. (2003). Treatment of experimental asthma by long-term gene therapy directed against IL-4 and IL-13. Mol. Ther. *7*, 155–162.

CHAPTER 5

Adeno-associated virus mediated gene therapy for vascular retinopathies

Brian J. Raisler,[1] Wen-Tao Deng,[2] Kenneth I. Berns[1] and William W. Hauswirth[1,2]

[1]*Department of Molecular Genetics and Microbiology, Center for Gene Therapy,*
[2]*Department of Ophthalmology, University of Florida, Gainesville, FL, USA*

5.1. Introduction

The vascular system in the retina of a healthy adult is quiescent with an intrinsic turnover rate of endothelial cells on the order of years. Damage to the vasculature as a result of trauma or diseases such as diabetes or atherosclerosis can render areas of the retina hypoxic. This leads to the growth of new vessels by angiogenesis, often referred to as neovascularization (NV) to differentiate it from developmental angiogenesis or collateral vessel formation. There is some evidence that vessels involved in NV can be distinguished from existing vessels. Endothelial precursor cell numbers are increased by pro-angiogenic factors that promote new capillary formation in the adult (Asahara et al., 1999). The process of NV can be differentiated from the process of prenatal vasculogenesis where cells known as hemangioblasts act as pluripotent progenitors capable of forming both blood and blood vessels (Noden, 1989; Choi, 1998). While the processes of angiogenesis and vasculogenesis each have a role in the formation and maintenance of the total vasculature, there are important differences. Vasculogenesis, primarily involved in developmental vessel formation, takes place by the de novo assembly of vasculature from endothelial precursors called angioblasts. Angiogenesis, involved both in developmental vessel formation and later processes such as wound healing, forms new micro-vessels by migration and proliferation of endothelial cells from larger, extant vessels. Although the two processes are somewhat distinct, evidence suggests that they are likely to share certain regulatory mechanisms. Lineage specific markers might be used to distinguish existing vasculature from pathologically formed angiogenic vessels, although a recent study suggests that adult hemapoetic stem cells may function as hemangioblasts (Grant et al., 2002) making such a distinction more difficult. Integrins $\alpha_v\beta_3$ and $\alpha_v\beta_5$ are present on endothelial cells participating in angiogenesis, but are absent on normal retinal endothelial cells (Friedlander et al., 1996; Luna et al., 1996). Regulation of vascularization in the mature retina involves a balance between endogenous positive growth factors, such as vascular endothelial growth factor (VEGF) (Ferrara and Davis-Smyth, 1997; Shweiki et al., 1992) and inhibitors of angiogenesis, such as pigment epithelium derived factor (PEDF) (Dawson et al., 1999). When this balance is upset, pathologic angiogenesis can occur, ultimately leading to a loss of vision. For a more complete review of angiogenesis and vasculogenesis, including the cell types and markers involved and the key regulatory players, interested readers should consult the review by Beck and D'Amore (1997).

Pathologic neovascularization of the retina is central to several debilitating ocular diseases including proliferative diabetic retinopathy (PDR), age-related macular degeneration (AMD), and retinopathy of prematurity (ROP). Diabetic retinopathy (primarily retinal NV) and the wet form of AMD (primarily choroidal NV) are the leading causes of blindness in developed countries.

5.1.1. Clinical treatments for retinal neovascular disease

Several clinical treatment options are currently available for patients presenting with retinal and choroidal neovascularization (CNV). Surgical, laser, and photodynamic therapy (for AMD) techniques have been most commonly used to treat both PDR and AMD (Spranger and Pfeiffer, 2001; Votruba and Gregor, 2001). Pan-retinal laser photo-coagulation has been used with relative success for a number of years (Le and Murphy, 1994; Boulanger et al., 2000). Unfortunately, laser photo-coagulation can also damage healthy retinal cells adjacent to or underlying the treated area leading to a significant loss of peripheral and night vision. Laser photo-coagulation for CNV secondary to AMD often leads to an immediate significant reduction in visual acuity (Lamkin and Singerman, 1994). A more recent development to treat CNV is photodynamic therapy that, through the use of vessel-targeted low energy lasers, reduces collateral damage during laser therapy (Smiddy and Flynn, 1999; Margherio et al., 2000; Schmidt-Erfurth and Hasan, 2000; Binder et al., 2002). Unfortunately, frequent re-treatment is necessary. Alternative treatments include surgery for removal of CNV membranes, foveal translocation, removal or displacement of subretinal blood for AMD patients, and removal of vitreous hemorrhage and scarring secondary to fibrovascular proliferation in PDR patients (Dogru et al., 1999; Ladd et al., 2001; Luke et al., 2001). All these surgical interventions carry intrinsic risks to the patient and can create further complications (Spranger and Pfeiffer, 2001; Votruba and Gregor, 2001). In addition, all current treatment options provide solutions that result in some loss of vision and are frequently only temporary. Recurrence of symptoms is common and often ultimately results in loss of vision. Clearly the need exists for therapies that require minimal surgical manipulation, preserve existing vision, and provide long-term amelioration of any future of NV.

5.2. New strategies for treating NV

5.2.1. VEGF and the hypoxia signaling pathway

Vascular endothelial growth factor (VEGF) appears to be the major pro-angiogenic factor in the retina and is induced through a hypoxia signaling pathway. A cytosolic heme protein appears to act as a sensor to detect decreased oxygen tension and to generate free radicals. This process, in turn, activates various transcription factors (Bunn and Poyton, 1996) including hypoxia inducible factor (HIF-1). HIF-1 stimulates transcription of multiple genes in response to hypoxia, including VEGF (Forsythe et al., 1996; Gerber et al., 1997). The stimulatory mechanism involves HIF-1 binding to a hypoxia response element (HRE) in the promoter region of the VEGF receptor-1 (Iyer et al., 1998). This suggests that HIF-1 signaling plays a key role in VEGF mediated neovascularization in response to local retinal ischemia. VEGF acts as a major angiogenic stimulator early in the signaling cascade, is clearly involved in retinal NV, and may also have a role in choroidal NV. Not surprisingly, therefore, several new approaches to control retinal NV involve attempts to modulate the VEGF induction pathway or VEGF directly by interfering with its angiogenic stimulation.

Current treatment options for patients with ocular neovascularization do not include antiangiogenic treatments, but such approaches are on the horizon. Recent reports have examined orally active drugs that inhibit the VEGF receptor kinase pathway and found them to significantly reduce ocular neovascularization in mice (Aiello et al., 1997; Bold et al., 2000). Systemic inhibition of the VEGF pathway of angiogenesis raises safety concerns that must be addressed before this approach could be applied clinically. To avoid concerns related to systemic angiogenic inhibition, local delivery of several agents is being investigated. Intravitreous injection of either soluble VEGF receptors or antisense oligonucleotides for VEGF reduce the retinal NV in a mouse model of ischemic retinopathy (Aiello et al., 1995; Robinson et al., 1996). Work in non-human primates has shown a reduction in NV of the iris following intravitreous injection of anti-VEGF antibody (Adamis et al., 1996). Recently, intraocular injections of a Fab fragment of an anti-VEGF antibody or an aptomer that binds VEGF have been tested for safety in phase II clinical trials for treatment of cancer and phase III trials both for cancer and control of angiogenesis in AMD are being planned (Ferrara, 2002). Preliminary reports suggest that injection of the anti-VEGF antibody may induce a local inflammatory response, but it is not considered a severe enough problem to discontinue these approaches (Ferrara and Gerber, 2001). Another promising avenue of reducing the neovascular response is intraocular injection of endogenous proteins with known antiangiogenic properties. Endogenous proteins are likely to be better tolerated than foreign antibodies or aptomers. Several alternative proteins with purported antiangiogenic activity have been identified (Taraboletti et al., 1990; Tombran-Tink et al., 1991; Cao, 1998), and intraocular administration of each of these alone or in combination could also be considered. All of these treatments, however, share the potentially limiting disadvantage of requiring repeated intraocular injection.

5.2.2. Gene-based therapies

The use of gene based therapy for the treatment of ocular neovascular disease offers advantages over conventional methods. By using a viral vector with a selective promoter to express the antiangiogenic protein or factor locally, expression can be limited to a specific cell type or subset of cell types within the retina. This reduces the safety concerns relative to systemic administration of antiangiogenic agents. Delivery of the vector to discreet compartments within the eye by sub-retinal or intravitreous injection may allow additional control of expression to only those local vessels that are affected. Choice of the appropriate viral vector for delivery of the therapeutic gene allows modulation of the duration of expression. Adeno-associated viral vectors appear to provide extended, perhaps even life-long, expression of therapeutic proteins within the eye. This may obviate the need for repeated intraocular injections in the case of chronic NV disease such as PDR or AMD. One conceivable treatment scenario is that patients seen for surgical or laser treatment of active neovascularization could receive concurrent vector treatment that would provide long-term protection from subsequent neovascular events.

In addition to the work presented here, several alternative viral-vectored approaches have been reported recently. An adeno-associated viral vector (AAV) encoding the soluble VEGF receptor 1, sFlt-1, shows promise for long-term inhibition of two types of ocular neovascularization (Lai et al., 2002). This vector, when injected into the anterior chamber, resulted in expression in both the corneal endothelium and iris pigment epithelium and reduced corneal NV by 36%. Subretinal injection of the same vector reduced choroidal NV subsequent to laser lesions around the optic nerve. These results suggest that a secretable factor expressed in one or more transduced cell populations can be effective in the control of ocular NV occurring in a disparate cell population.

Another gene-based approach to treating neovascular disease involves the virally mediated intraocular expression of antiangiogenic agents. Several factors that have been identified as potential angiogenic inhibitors could be delivered in a viral-vectored system rather than by direct injection. By expressing the angiogenic inhibitor directly where it is needed and for a longer duration than a single protein injection, greater efficacy in the control of disease may be possible. In addition to our studies, this strategy was tested by expressing several antiangiogenic factors from AAV vectors: Tissue inhibitor of metalloproteinase-3 (TIMP3), pigment epithelium derived factor (PEDF), and endostatin (Auricchio et al., 2002). Each of these factors successfully reduced the level of neovascularization in an ischemic mouse model of retinopathy of prematurity. A similar strategy is also being tested in a phase I clinical trial using adenovirus to deliver PEDF as a potential treatment for wet AMD (Rasmussen et al., 2001). Adenovirus vectors have a shorter expression period than that mediated by AAV, a difference that might be exploited depending on the exogenous factors being expressed and the target ocular disease.

Two of the most potent general inhibitors of neovascularization are Kringle domains 1 through 3 of angiostatin (K1K3) and pigment epithelium derived factor (PEDF). Angiostatin is a proteolytic fragment of plasminogen and its K1 to K3 domain retains potent angiostatic properties. It is an endogenous regulator of vasculogenesis and, as a naturally occurring peptide, it is not likely to stimulate an immunogenic response (Cao, 2001; Kirsch et al., 2000). Neither plasminogen nor plasmin inhibits endothelial cell proliferation, nor does angiostatin affect coagulation. Although angiostatin is known to inhibit tumor growth in vivo by increasing apoptosis and inhibiting tumor associated angiogenesis, its precise mechanism of action is unclear. Apoptosis in vitro is induced in endothelial cells by multiple forms of angiostatin (Lucas et al., 1998) and cells have been shown to be arrested at the G2/M transition interface (Griscelli et al., 1998). Administration of angiostatin to tumor bearing mice has not resulted in detectable systemic cytotoxicity; only angiogenic proliferation is inhibited (O'Reilly et al., 1994, 1996; Wu et al., 1997). Angiostatin therefore appears to be an effective and non-toxic inhibitor of NV that is worth evaluating in models of ischemic retinopathy. A recent study has indicated that Kringle 5 of angiostatin may induce PEDF and inhibit VEGF both in cell culture and a rat model of ischemic retinopathy (Gao et al., 2002).

PEDF, first purified from human retinal pigment epithelial cultures as a factor that induces neuronal differentiation of cultured retinoblastoma cells (Tombran-Tink et al., 1991; Steele et al., 1993), has been recently shown to regulate normal angiogenesis in the eye (Dawson et al., 1999). PEDF is found both intracellularly and extracellularly in the fetal and early adult eye but is lost at the onset of senescence (Becerra, 1997; Araki et al., 1998). It is down-regulated by hypoxia and induced in the retina as a result of hyperoxia; it is a very potent inhibitor of corneal NV and prevents endothelial cell migration towards a wide variety of angiogenic inducers (Dawson et al., 1999). PEDF therefore appears to be a major endogenous angiogenic regulator of the retinal vasculature and is an excellent candidate gene for therapy against pathogenic ocular NV. As an intraocularly injected protein, PEDF has been shown to delay the loss of photoreceptors in the retinal degeneration (rd) mouse (Cayouette et al., 1999), implying that it may also possess neurotrophic activity in the retina and that the extracellular protein can effectively disperse throughout the retina. We hypothesize here that when expressed in a secretable form as a viral-vectored cDNA, either PEDF or K1K3 may act relatively independent of the retinal cell type supporting expression and may be effective in limiting retinal or choroidal NV.

In order to determine whether either or both PEDF or K1K3 are potentially useful for therapeutic control of retinal NV, we examined the effect of expression of the angiostatic factors PEDF and K1K3 in a mouse model of ischemic retinopathy. We also investigated PEDF for control of choroidal NV in a laser-induced mouse model (Mori et al., 2002), but will focus here on the former studies. We chose recombinant adeno-associated virus (rAAV) serotype-2 to deliver the therapeutic genes because issues regarding attainment of high vector titers have been resolved (Hauswirth et al., 2000) and rAAV preparations free of contaminating replication competent rAAV are now routine (Robinson et al., 1996; Zolotukhin et al., 1999; Hauswirth et al., 2000). Recombinant AAV mediated gene delivery results in long-term expression in a wide variety of tissues, including various cell types in the retina (Flannery et al., 1997) and optic nerve.

Although the minimum effective intraocular dose of PEDF or K1K3 for controlling ischemic retinopathy is unknown, previous studies following systemic administration of PEDF protein in the same model we employ here suggest that the therapeutic threshold is approximately 5-11 μg/day (Stellmach et al., 2001) by I.P. administration. If we make the simplest and most conservative assumptions, that PEDF is stable and partitioned into the eye from the systemic vasculature based simply on the volume of the eye versus the volume of the whole animal, the threshold concentration of ocular PEDF needed to inhibit retinal NV is estimated to be 1–2 ng/eye. This conservative calculation does not consider any pharmacokinetic stability parameters or the existence of the blood-retinal barrier, both of which would tend to reduce further the amount of PEDF reaching the eye from the circulation. It is therefore possible that the true intraocular therapeutic threshold for PEDF is much lower than this rough estimate. We have demonstrated that ocularly administered rAAV vector expressing PEDF produces intraocular concentrations of 20–70 ng in the adult mouse and 1–8 ng in the neonatal mouse (Raisler et al., 2002). This is at or considerably higher than our estimated therapeutic threshold. Similar calculations based on another study in which one dose of K1K3 angiostatin administered systemically was able to effectively reduce retinal NV in the ROP mouse (Meneses et al., 2001) yield an estimated maximum intraocular therapeutic threshold of 1.6 ng/eye for K1K3. Our rAAV vector expressing K1K3 angiostatin produces levels of 6–60 ng/eye in the adult or 2–3 ng/eye in the neonate again above our estimated maximum threshold.

Therefore it appears that we can achieve and maintain intraocular levels of either PEDF or K1K3 angiostatin in neonatal and adult mice sufficient to expect significant reduction of retinal NV. In the ischemic mouse model the level of retinal NV is measured quantitatively by enumerating the endothelial cells above the inner limiting membrane (ILM) of the retina (see later). Such an analysis showed that PEDF treated eyes had 74% fewer endothelial cells above the ILM compared to paired controls and 78% fewer compared to paired controls for K1K3 treated eyes (Raisler et al., 2002).

Administration of rAAV vectors expressing cDNAs for antiangiogenic proteins can therefore be as effective as administration of the corresponding proteins systemically, but without the requirement for repeated injections. In addition, local production and secretion of PEDF or K1K3 through vector gene delivery is likely to restrict any antineovascular activity to an area specific to the pathological angiogenesis in the subject. Proteins produced in the posterior ocular compartment through viral vectors are not likely to interfere with the normal angiogenic processes necessary for wound healing or tissue repair elsewhere in the body, and perhaps not even in the anterior segment of the same eye, although this remains to be tested. Furthermore, rAAV vectors have demonstrated long-term, sustained high-level expression in the retina (Guy et al., 1999), and we observe nanogram levels in rat eyes for at least 21 months (B. Raisler and W. Hauswirth, unpublished), indicating that a single injection of rAAV bearing a therapeutic gene could provide durable inhibition of NV. This could obviate the need for repeated injections of antiangiogenic compounds systemically or intraocularly.

For the purposes of maximizing protein expression in the retina, we found that rAAV-CBA-PEDF (chicken β-actin promoter regulating PEDF mRNA synthesis) and rAAV-CBA-K1K3 vectors produced the highest ocular levels of protein. The lower levels of expression seen for vectors with mouse opsin (MOPS), cis-retinaldehyde binding protein (CRALBP), and platelet-derived growth factor (PDGF) promoters may be due to the different cell specificity of these promoters compared to CBA or to an attenuated ability to support transcription in the same set of retinal cell types. When subretinally injected, CBA-containing vectors support expression well in both RPE cells and photoreceptors (Acland et al., 2001). In contrast, the MOPS promoter is rod photoreceptor specific (Flannery et al., 1997) and yields much lower intraocular expression, thus suggesting that the key cellular source of secreted PEDF or K1K3 from vectors injected subretinally is the RPE cell. When injected into the vitreous, CBA-containing rAAV vectors express very well, but now predominantly in retinal ganglion cells (RGC's) (Mori et al., 2002). Since there was no significant difference in the antineovascular effectiveness of subretinally or intravitreally administered PEDF or K1K3 vectors with the CBA promoter, it appears that the potentially less traumatic intravitreal route of vector delivery may be favored. For optimizing the safety of ocular gene therapy for NV diseases it may be important to limit expression of a therapeutic protein even more specifically, either to just a single retinal cell type or to a more defined topologic area. By altering the promoter used to drive PEDF or K1K3 expression it may be possible to fine-tune therapeutic gene expression to a pharmacologically significant but highly localized cellular pattern. The CBA promoter drives expression in multiple cell types whereas a more specific promoter could in theory be employed to target expression with more cell-selectivity. Alternatively, delivering the vector specifically to the intravitreal or subretinal space in a larger human eye may define a relatively local region of expression. Advanced stages of AMD are characterized by a neovascularization of the choriocapillaris within or adjacent to the macula. In this case, treatment might be most effective if the therapeutic vector is administered subretinally near or surrounding potentially active CNV regions. This type of subretinal administration might limit the lateral spread of vector-mediated expression (Acland et al., 2001) relative to vitreal administration that allows less constrained vector diffusion to a wider and hence less controlled retinal area. Full testing of these ideas will require development of an animal model with a retinal area closer in size to that in humans than the mouse.

In summary, we have demonstrated that rAAV vectors incorporating a CBA promoter are capable of producing sustained therapeutic levels of PEDF and K1K3 in the mouse eye. Intraocular injection of either rAAV-CBA-PEDF or rAAV-CBA-K1K3 significantly reduced the level of retinal NV in a mouse model of ischemic retinopathy. Independent studies have recently demonstrated similarly effective gene therapy approaches for controlling ocular angiogenesis. Expression of PEDF from either rAAV (Mori et al., 2002) or adenovirus (Mori et al., 2001b) vectors were effective in reducing choroidal NV in rodent models. Adenoviral vectors expressing endostatin (Mori et al., 2001a) or plasminogen activator inhibitor-1 (Lambert et al., 2001) were also effective in inhibiting retinal NV. Together with our present report, the generality of efficient and well-targeted gene-based approaches for treating neovascular diseases of the eye coupled with the potential of rAAV vectors for persistently delivering sustained therapeutic levels of antiangiogenic proteins to the retina is becoming apparent. Finally, rAAV vectors have shown very little in vivo toxicity in a variety of tissues (Song et al., 1998).

5.3. Protocols

5.3.1. Injection of viral vectors

1. Neonatal mouse pups, obtained within less than 16 h after birth, are injected in one eye with the therapeutic viral vector. The contralateral eye serves as a control and may remain uninjected, be mock-injected with carrier solution, or be injected with a non-therapeutic vector as the experiment indicates. Having determined that mock-injection or non-therapeutic injection did not impact the experimental model, our subsequent control eyes were uninjected.
2. The pups are sedated by hypothermia until their motor activity is slowed.
3. Mix the virus with a fluorescein dye (0.1 mg/ml final concentration), to better visualize the injection.
4. The virus is injected using a 2.5 μl-capacity syringe with a removable 33 gauge, 30° bevel needle (Hamilton Company, Reno, NV). A quantity of 0.5 μl is delivered intraocularly in each treated eye. Under a dissecting microscope, the 33-gauge needle penetrates the fused skin over the eye and then the cornea of the eye. To determine the depth of penetration it is best to angle the bevel of the needle up. By carefully advancing the needle until the bevel is fully hidden from view and then advancing half again this distance it is possible to reproducibly inject vector into the vitreal space in neonatal pup eyes. If dye can be seen under the skin, the injection was too shallow

and never penetrated the eye. A good injection should show no dye under the skin in pigmented animals indicating the virus is inside the pigmented eye.

5.3.2. *Hyperoxia treatment*

1. Mouse pups, with their nursing dam, are placed in a chamber at 73% oxygen at P7 and maintained in this environment for 5 days until P12.
2. At P12, the pups and nursing dam are returned to normal room air and maintained for another 5 days.
3. At P17, the pups are euthanized and their eyes enucleated and fixed for embedding and sectioning. Representative pups from each group are anesthetized and perfused through the left ventricle with 4% paraformaldehyde in 0.1 M sodium phosphate (pH 7.4) containing FITC-dextran for visualizing the retinal vasculature (see later).

5.3.3. *Qualitative assessment of retinal NV*

For qualitative assessment of retinal NV, both eyes of each perfused P17 pup are enucleated and the retina dissected and flat-mounted as described by D'Amato et al. (1993).

1. Anesthetize the animals and perfuse through the left ventricle with 4% paraformaldehyde in 0.1 M sodium phosphate (pH 7.4) containing 5 mg/ml FITC-dextran (2,000,000 molecular weight, Sigma-Aldrich, St. Louis, MO).
2. Enucleate the eyes, puncture through the cornea with a 27-gauge needle, and fix in the paraformaldehyde solution without the FITC label for 30–60 min.
3. Wash the eyes 3 times for 15 min in phosphate buffered saline (PBS).

4. Using a fresh, sharp scalpel blade for each new eye, cut off the cornea from the eyecup just in front of the corneal–scleral junction.

5. With the cornea removed, work the retina away from the eye cup by sqeezing gently on the scleral surface near the optic nerve and teasing the retina out with a fine haired brush. It is best to do this step with the eyecup sitting a small pool of PBS. This minimizes any tearing of the retina and makes handling easier.

6. Place the retina, with the interior surface facing up so that it resembles a small cup, in a pool of PBS. Using a sharp flat scalpel blade make 4–5 radial cuts around the perimeter of the retina. These cuts should extend about 1/2 to 2/3 the distance from the edge to the center of the retina and should allow the retina to lay flat on the slide. The retinal edges may have to be gently unfolded with the fine hair brush to obtain a flat retina.

7. Add a drop of Fluor-Mount (Southern Biotech, Birmingham, AL) to delay quenching of the FITC signal in the retinal vessels, and then add the coverslip avoiding bubbles.

8. Flatmounted retinas are photographed by fluorescence microscopy using a Zeiss Axioplan2 microscope, Zeiss Plan-Fluar 10× lens, Sony DXC–970MD camera with tile field imaging and MCID software (Imaging Research Inc.). At least three eyes from each treatment group should be examined in this way to gain an accurate qualitative assessment of retinal vessel patterns.

5.3.4. Quantitative assessment of retinal NV

Both eyes of each P17 pup are enucleated and fixed for paraffin embedding and serially sectioned at 5 μm thickness as described by Smith et al. (1994).

1. Representative sections (every 30th section) through the full eyecup are stained with hematoxylin and eosin to visualize cell nuclei.

2. Trained investigators masked to the identity of the treatment groups count neovascular endothelial cell nuclei above the internal limiting membrane in every 30th section through each eye. These representative sections provide a basis for comparing the effect of expression of therapeutic factors without the need for enumerating every section through the eye. In sections containing any portion of the optic nerve, endothelial cell nuclei directly above or associated with the optic nerve head are not counted as neovascular nuclei. Any endothelial vessels above the optic nerve head or within one ONH diameter distance from it may represent persistent hyaloid vessels and should not be included in the data analysis (Fig. 5.1).

3. Vascular cell nuclei are considered to be associated with neovascularization if they are on the vitreous side of the internal limiting membrane. Only endothelial cell nuclei that are clearly part of a vessel structure with a discernable lumen in any plane are included in our analysis (Fig. 5.2). All such cells are counted unless

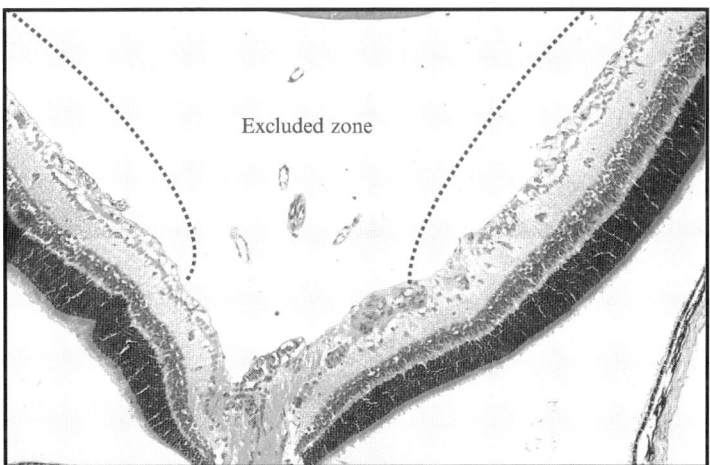

Fig. 5.1. Retinal section near the optic nerve head showing persistent hyaloid vessels in an ROP mouse. Such regions are excluded from retinal vascular endothelial cell enumeration.

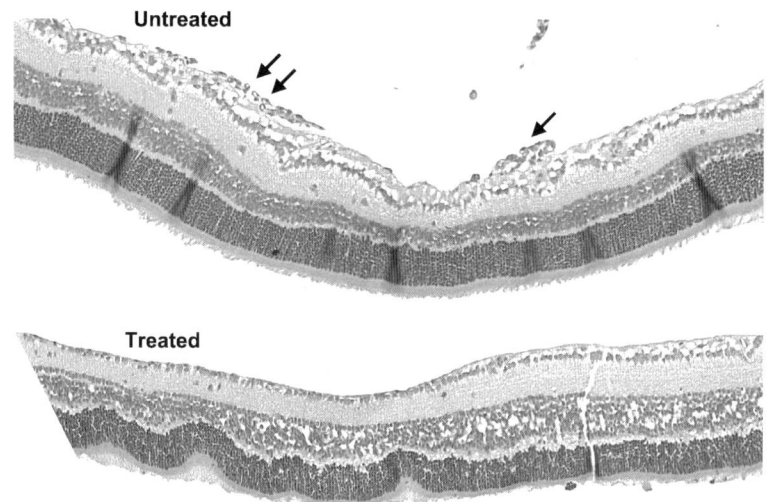

Fig. 5.2. Retinal cross-sections of P17 mice exposed to hyperoxia in the ROP protocol. Significant numbers of neovascular tufts are apparent on the vitreal side of the ILM in untreated mice (top image). Two large vessels protruding beyond the ILM are indicated by arrows in the untreated eye. Many fewer such nuclei were observed in vector treated eyes (bottom image).

they were clearly associated with hyaloid structures as noted above. Single nuclei or clumps of nuclei not defining a vessel are not counted as they may represent sectioning artifacts.

4. Data are analyzed by a paired t-test with vector treated and contralateral uninjected eyes serving as determinants.

5.4. Discussion

Effective, long lasting treatment of retinal neovascular disorders, including proliferative diabetic retinopathy and proliferative AMD, remains one of the greatest challenges in ophthalmology today. The number of individuals suffering from diabetes has increased worldwide in recent years and is projected to continue to rise (Mokdad et al., 2001). PDR is a common complication in diabetic patients. PDR shares a patho-physiology with our model of ischemic retinopathy since the initial ischemic insult and the subsequent pathologic outgrowth of new vessels from the retinal vasculature occurs in both PDR and ROP and ultimately leads to blindness in either disease. Retinal NV can also occur secondarily to choroidal NV in AMD (Lip et al., 2001; Yannuzzi et al., 2001).

Administration of rAAV vectors expressing cDNAs for antiangiogenic proteins can therefore be as effective as administration of the corresponding proteins systemically, but without the requirement for repeated injections. In addition, use of rAAV vectors offers several potentially important advantages over systemic administration of antiangiogenics. Local production and secretion of PEDF or K1K3 through vector gene delivery is likely to restrict any antineovascular activity to an area specific to the pathological angiogenesis in the subject. Proteins produced in the posterior ocular compartment are not likely to interfere with the normal angiogenic processes necessary for wound healing or tissue repair elsewhere in the body, and perhaps not even in the anterior segment of the same eye, although this remains to be tested. Further, rAAV vectors have demonstrated long-term, sustained high-level expression in the retina (Guy et al., 1999), and we observe nanogram levels in rat eyes for at least 21 months (B. Raisler and W. Hauswirth, unpublished), indicating that a single injection of rAAV bearing a therapeutic gene could provide durable inhibition of NV. This would obviate the need for repeated injections of antiangiogenic compounds systemically or intraocularly.

References

Acland, G. M., Aguirre, G. D., Ray, J., Zhang, Q., Aleman, T. S., Cideciyan, A. V., Pearce-Kelling, S. E., Anand, V., Zeng, Y. Maguire, A. M. et al. (2001). Nat. Genet. *28*, 92–95.

Adamis, A. P., Shima, D. T., Tolentino, M. J., Gragoudas, E. S., Ferrara, N., Folkman, J., D' Amore, P. A. and Miller, J. W. (1996). Arch. Ophthalmol. *114*, 66–71.

Aiello, L. P., Bursell, S. E., Clermont, A., Duh, E., Ishii, H., Takagi, C., Mori, F., Ciulla, T. A., Ways, K. Jirousek, M. et al. (1997). Diabetes *46*, 1473–1480.

Aiello, L. P., Pierce, E. A., Foley, E. D., Takagi, H., Chen, H., Riddle, L., Ferrara, N., King, G. L. and Smith, L. E. (1995). Proc. Natl. Acad. Sci. USA *92*, 10457–10461.

Araki, T., Taniwaki, T., Becerra, S. P., Chader, G. J. and Schwartz, J. P. (1998). J. Neurosci. Res. *53*, 7–15.

Asahara, T., Masuda, H., Takahashi, T., Kalka, C., Pastore, C., Silver, M., Kearne, M., Magner, M. and Isner, J. M. (1999). Circ. Res. *85*, 221–228.

Auricchio, A., Behling, K. C., Maguire, A. M., O' Connor, E. M., Bennett, J., Wilson, J. M. and Tolentino, M. J. (2002). Mol. Ther. *6*, 490–494.

Becerra, S. P. (1997). Adv. Exp. Med. Biol. *425*, 223–237.

Beck, L., Jr. and D'Amore, P. A. (1997). FASEB J. *11*, 365–373.

Binder, S., Stolba, U., Krebs, I., Kellner, L., Jahn, C., Feichtinger, H., Povelka, M., Frohner, U., Kruger, A. Hilgers, R. D. et al. (2002). Am. J. Ophthalmol. *133*, 215–225.

Bold, G., Altmann, K. H., Frei, J., Lang, M., Manley, P. W., Traxler, P., Wietfeld, B., Bruggen, J., Buchdunger, E. Cozens, R. et al. (2000). J. Med. Chem. *43*, 2310–2323.

Boulanger, A., Liu, S., Henningsgaard, A. A., Yu, S. and Redmond, T. M. (2000). J. Biol. Chem. *275*, 31274–31282.

Bunn, H. F. and Poyton, R. O. (1996). Physiol Rev. *76*, 839–885.

Cao, Y. (1998). Prog. Mol. Subcell. Biol. *20*, 161–176.

Cao, Y. (2001). Int. J. Biochem. Cell Biol. *33*, 357–369.

Cayouette, M., Smith, S. B., Becerra, S. P. and Gravel, C. (1999). Neurobiol. Dis. *6*, 523–532.

Choi, K. (1998). Biochem. Cell Biol. *76*, 947–956.

D'Amato, R., Wesolowski, E. and Smith, L. E. (1993). Microvasc. Res. *46*, 135–142.

Dawson, D. W., Volpert, O. V., Gillis, P., Crawford, S. E., Xu, H., Benedict, W. and Bouck, N. P. (1999). Science *285*, 245–248.

Dogru, M., Nakamura, M., Inoue, M. and Yamamoto, M. (1999). Jpn. J. Ophthalmol. *43*, 217–224.

Ferrara, N. (2002). Semin. Oncol. *29*, 10–14.

Ferrara, N. and Davis-Smyth, T. (1997). Endocr. Rev. *18*, 4–25.

Ferrara, N. and Gerber, H. P. (2001). Acta Haematol. *106*, 148–156.

Flannery, J. G., Zolotukhin, S., Vaquero, M. I., LaVail, M. M., Muzyczka, N. and Hauswirth, W. W. (1997). Proc. Natl. Acad. Sci. USA *94*, 6916–6921.
Forsythe, J. A., Jiang, B. H., Iyer, N. V., Agani, F., Leung, S. W., Koos, R. D. and Semenza, G. L. (1996). Mol. Cell Biol. *16*, 4604–4613.
Friedlander, M., Theesfeld, C. L., Sugita, M., Fruttiger, M., Thomas, M. A., Chang, S. and Cheresh, D. A. (1996). Proc. Natl. Acad. Sci. USA *93*, 9764–9769.
Gao, G., Li, Y., Gee, S., Dudley, A., Fant, J., Crosson, C. and Ma, J. X. (2002). J. Biol. Chem. *277*, 9492–9497.
Gerber, H. P., Condorelli, F., Park, J. and Ferrara, N. (1997). J. Biol. Chem. *272*, 23659–23667.
Grant, M. B., May, W. S., Caballero, S., Brown, G. A., Guthrie, S. M., Mames, R. N., Byrne, B. J., Vaught, T., Spoerri, P. E. Peck, A. B. et al. (2002). Nat. Med. *8*, 607–612.
Griscelli, F., Li, H., Bennaceur-Griscelli, A., Soria, J., Opolon, P., Soria, C., Perricaudet, M., Yeh, P. and Lu, H. (1998). Proc. Natl. Acad. Sci. USA *95*, 6367–6372.
Guy, J., Qi, X., Muzyczka, N. and Hauswirth, W. W. (1999). Arch. Ophthalmol. *117*, 929–937.
Hauswirth, W. W., Lewin, A. S., Zolotukhin, S., Muzyczka, N. and Palczewski, K. (2000). Methods Enzymol. 316.
Iyer, N. V., Kotch, L. E., Agani, F., Leung, S. W., Laughner, E., Wenger, R. H., Gassmann, M., Gearhart, J. D., Lawler, A. M. Yu, A. Y. et al. (1998). Genes Dev. *12*, 149–162.
Kirsch, M., Schackert, G. and Black, P. M. (2000). J. Neurooncol. *50*, 173–180.
Ladd, B. S., Solomon, S. D., Bressler, N. M. and Bressler, S. B. (2001). Am. J. Ophthalmol. *132*, 659–667.
Lai, Y. K., Shen, W. Y., Brankov, M., Lai, C. M., Constable, I. J. and Rakoczy, P. E. (2002). Gene Ther. *9*, 804–813.
Lambert, V., Munaut, C., Noel, A., Frankenne, F., Bajou, K., Gerard, R., Carmeliet, P., Defresne, M. P., Foidart, J. M. and Rakic, J. M. (2001). FASEB J. *15*, 1021–1027.
Lamkin, J. C. and Singerman, L. J. (1994). Semin. Ophthalmol. *9*, 10–22.
Le, D. and Murphy, R. P. (1994). Semin. Ophthalmol. *9*, 2–9.
Lip, P. L., Blann, A. D., Hope-Ross, M., Gibson, J. M. and Lip, G. Y. (2001). Ophthalmology *108*, 705–710.
Lucas, R., Holmgren, L., Garcia, I., Jimenez, B., Mandriota, S. J., Borlat, F., Sim, B. K., Wu, Z., Grau, G. E. Shing, Y. et al. (1998). Blood *92*, 4730–4741.

Luke, C., Aisenbrey, S., Luke, M., Marzella, G., Bartz-Schmidt, K. U. and Walter, P. (2001). Br. J. Ophthalmol. *85*, 928–932.

Luna, J., Tobe, T., Mousa, S. A., Reilly, T. M. and Campochiaro, P. A. (1996). Lab Invest *75*, 563–573.

Margherio, R. R., Margherio, A. R. and De Santis, M. E. (2000). Retina *20*, 325–330.

Meneses, P. I., Hajjar, K. A., Berns, K. I. and Duvoisin, R. M. (2001). Gene Ther. *8*, 646–648.

Mokdad, A. H., Ford, E. S., Bowman, B. A., Nelson, D. E., Engelgau, M. M., Vinicor, F. and Marks, J. S. (2001). Diabetes Care *24*, 412.

Mori, K., Ando, A., Gehlbach, P., Nesbitt, D., Takahashi, K., Goldsteen, D., Penn, M., Chen, C. T., Mori, K. Melia, M. et al. (2001a). Am. J. Pathol. *159*, 313–320.

Mori, K., Duh, E., Gehlbach, P., Ando, A., Takahashi, K., Pearlman, J., Mori, K., Yang, H. S., Zack, D. J. Ettyreddy, D. et al. (2001b). J. Cell Physiol. *188*, 253–263.

Mori, K., Gehlbach, P., Yamamoto, S., Duh, E., Zack, D. J., Li, Q., Berns, K. I., Raisler, B. J., Hauswirth, W. W. and Campochiaro, P. A. (2002). Invest Ophthalmol. Vis. Sci. *43*, 1994–2000.

Noden, D. M. (1989). Am. Rev. Respir. Dis. *140*, 1097–1103.

O'Reilly, M. S., Holmgren, L., Chen, C. and Folkman, J. (1996). Nat. Med. *2*, 689–692.

O'Reilly, M. S., Holmgren, L., Shing, Y., Chen, C., Rosenthal, R. A., Moses, M., Lane, W. S., Cao, Y., Sage, E. H. and Folkman, J. (1994). Cell *79*, 315–328.

Raisler, B. J., Berns, K. I., Grant, M. B., Beliaev, D. and Hauswirth, W. W. (2002). Proc. Natl. Acad. Sci. USA *99*, 8909–8914.

Rasmussen, H., Chu, K. W., Campochiaro, P., Gehlbach, P. L., Haller, J. A., Handa, J. T., Nguyen, Q. D. and Sung, J. U. (2001). Hum. Gene Ther. *12*, 2029–2032.

Robinson, G. S., Pierce, E. A., Rook, S. L., Foley, E., Webb, R. and Smith, L. E. (1996). Proc. Natl. Acad. Sci. USA *93*, 4851–4856.

Schmidt-Erfurth, U. and Hasan, T. (2000). Surv. Ophthalmol. *45*, 195–214.

Shweiki, D., Itin, A., Soffer, D. and Keshet, E. (1992). Nature *359*, 843–845.

Smiddy, W. E. and Flynn, H. W., Jr. (1999). Surv. Ophthalmol. *43*, 491–507.

Smith, L. E., Wesolowski, E., McLellan, A., Kostyk, S. K., D'Amato, R., Sullivan, R. and D'Amore, P. A. (1994). Invest Ophthalmol. Vis. Sci. *35*, 101–111.

Song, S., Morgan, M., Ellis, T., Poirier, A., Chesnut, K., Wang, J., Brantly, M., Muzyczka, N., Byrne, B. J. Atkinson, M. et al. (1998). Proc. Natl. Acad. Sci. USA *95*, 14384–14388.

Spranger, J. and Pfeiffer, A. F. (2001). Exp. Clin. Endocrinol. Diabetes *109*, S438–S450.

Steele, F. R., Chader, G. J., Johnson, L. V. and Tombran-Tink, J. (1993). Proc. Natl. Acad. Sci. USA *90*, 1526–1530.

Stellmach, V., Crawford, S. E., Zhou, W. and Bouck, N. (2001). Proc. Natl. Acad. Sci. USA *98*, 2593–2597.

Taraboletti, G., Roberts, D., Liotta, L. A. and Giavazzi, R. (1990). J. Cell Biol. *111*, 765–772.

Tombran-Tink, J., Chader, G. G. and Johnson, L. V. (1991). Exp. Eye Res. *53*, 411–414.

Votruba, M. and Gregor, Z. (2001). Eye *15*, 424–429.

Wu, Z., O'Reilly, M. S., Folkman, J. and Shing, Y. (1997). Biochem. Biophys. Res. Commun. *236*, 651–654.

Yannuzzi, L. A., Negrao, S., Iida, T., Carvalho, C., Rodriguez-Coleman, H., Slakter, J., Freund, K. B., Sorenson, J., Orlock, D. and Borodoker, N. (2001). Retina *21*, 416–434.

Zolotukhin, S., Byrne, B. J., Mason, E., Zolotukhin, I., Potter, M., Chesnut, K., Summerford, C., Samulski, R. J. and Muzyczka, N. (1999). Gene Ther. *6*, 973–985.

CHAPTER 6

Gene therapy for prevention and treatment of type 1 diabetes

Matthias H. Kapturczak,[1] Brant R. Burkhardt[2] and Mark A. Atkinson[3]

[1]*Department of Medicine, University of Alabama at Birmingham, 1530 3rd Avenue South Birmingham, AL 35294-0007, USA*
[2]*Department of Pathology and Laboratory Medicine, Children's Hospital of Philadelphia, 3516 Civic Center Blvd. Philadelphia, PA 19104-4318, USA*
[3]*Department of Pathology, University of Florida, Box 100275, Gainesville, FL 32610-0275, USA*

6.1. The clinical problem diabetes

Type 1 diabetes (insulin-dependent diabetes or IDDM) presents a significant health problem in the US and all over the world. It affects as many as 1 in 300 people in the US leading to many health risks and complications including diabetic retinopathy, cardiovascular disease, neuropathy as well as nephropathy (Atkinson and Maclaren, 1994). As the human and financial burden associated with this disease is enormous, much effort has been directed towards development of adequate treatment measures geared for a tight control of blood glucose levels. Despite major advances in insulin replacement therapy, including insulin pumps with incorporated glucose-sensing mechanisms, the overall impact on reduction of diabetes-related health problems has been disappointing at best. Current approaches in fight against type 1 diabetes include, in general, design of preventive measures as well as development of methods for β cell replacement. The latter approach includes pancreas or pancreatic islet transplantation as well as islet cell neogenesis.

6.2. Transplantation

The transplantation of islet cells (either as a whole pancreas or isolated islets) has the potential of restoring the physiological regulation of insulin production afforded by β cells and represents an exciting alternative to exogenous insulin administration. It offers, when successful, excellent metabolic control in the absence of the severe hypoglycemic episodes associated with intensive insulin treatment (Hering et al., 1993; Kenyon et al., 1996; Rosenberg, 1998). An extensive research effort has therefore been undertaken to develop methods for preventing pancreatic β cell destruction following islet cell transplantation.

Clinical trials have established that long-term function of transplanted islets can be observed in selected recipients and that the mass of islets obtained from a single donor can be sufficient to obtain measurable function in the recipient (Scharp et al., 1991; Warnock et al., 1991; Alejandro et al., 1997). However, for a vast majority of individuals, islet transplantation remains unsuccessful, with a substantial percentage of recipients losing graft function within a short period of time. Several mechanisms are thought to concurrently contribute to this clinical islet graft failure. They include rejection, recurrence of anti-islet cell autoimmunity, and nonspecific islet loss immediately after transplantation due to perturbation of the graft microenvironment (i.e., inflammation, ischemia/reperfusion) (Bottino et al., 1998; Kaufman et al., 1990; Stevens et al., 1994; Weirs et al., 1990). For islet transplantation to become a clinical reality, a need exists to devise strategies of immunosuppression/immunomodulation that are substantially different from the currently utilized pharmacological immunosuppressive regimens. The latter approach is plagued by the apparent failure to reproducibly promote long-term islet graft survival, the direct toxic effects of these agents on islet function, and serious unwanted side effects linked to chronic immunosuppression (Dunn, 1990; Hering et al., 1993; Jindal, 1994; Kenyon et al., 1996; Masetti et al., 1997; Rosenberg, 1998).

In general, immunosuppressive treatment regimens can be linked to an increased incidence of infections, neoplasia, and to the acceleration of cardiovascular disease. Each specific group of anti-rejection pharmaceuticals has its own profile of metabolic and pathophysiological effects. Pancreatic β cells seem to be especially vulnerable in situations of an increased metabolic demand. The anti-rejection regimens currently utilized for both whole pancreas and islets transplantation are all based on steroids and calcinurin inhibitors. Unfortunately, conventional immunosuppressive agents routinely used in islet transplant patients (e.g., cyclosporine, tacrolimus, and steroids) are characterized by intrinsic diabetogenic effects imposing a two to threefold increase in the metabolic demand of islets (Hahn et al., 1986; Hirano et al., 1992). Although recent studies of novel experimental immunosuppressive regimens allowing for an excellent islet graft survival without the use of steroids have been reported (Shapiro et al., 2000), yet the consequences of long-term immunosuppression remain a significant threat. In short, the collective data to date clearly suggest that alternative approaches are needed to significantly improve the overall success of islet cell transplantation. In order to become successful, therapies must overcome the two major immune system based obstacles: Allograft rejection and recurrent autoimmunity.

6.3. Allograft rejection: mechanisms for increasing graft acceptance

For the foreseeable future, the source of choice for islet cell transplantation will be "allogeneic" pancreas donors. By definition, there is a disparity at the major histocompatibility complex

TABLE 6.1
Comparison of islet allograft destruction and recurrent autoimmunity in NOD mouse models

Characteristic	Allograft rejection	Recurrent autoimmunity
Disease		
Onset of rejection	2–3 days	14–28 days and beyond
Peak infiltrate	5–8 days	21 days
Disease onset	Fast	Moderate to slow
Pathology		
Macrophages	++	++
T cells	++++, $CD8^+$? $CD4^+$	++, $CD8^+ > CD4^+$
Cytokine balance	Predominantly Th1	Predominantly Th1
Target antigens	Allo (i.e., MHC)	Self (e.g., insulin)
Autoantibodies	Rare to absent	Often

(MHC) between the donor and recipient. The allogeneic rejection episodes differ from autoimmune processes leading to β cell destruction in type 1 diabetes (Table 6.1). Without pharmacological intervention involving immunosuppressive drugs, the host's immune response will reject such an "allogeneic" or "non-self" graft, predominantly by actions involving T cells. Early education of T cells to recognize non-self islets, an obviously unnatural situation in nature, occur within the thymus. The state known as "tolerance", the inability of self-T cells to react against self-tissues, is acquired as immature T cells develop and pass through this organ [reviewed in Pleyer et al. (2000)]. The failure to react against one's self-tissues occurs because a majority of potentially autoreactive T cells are negatively selected by clonal deletion, with clonal anergy (i.e., antigen non-responsiveness) and regulatory T cells playing a role in regulation of responses in the peripheral immune system compartment.

To design a method for inducing long-term tolerance to alloantigens in graft recipients is an important goal for transplant immunologists (Pleyer et al., 2000). A state of specific tolerance would allow recipients to tolerate (i.e., not reject) the transplanted organ, and still be able to respond to other foreign antigens (e.g., viruses, bacteria). A number of therapeutic interventions like the adoptive cotransfer of bone marrow cells as well as interference with T cell costimulatory signals (e.g., anti-CD154) hold promise in this regard (Kenyon et al., 1999). Gene therapy has been considered to have a great potential as a complementary, if not exclusive, approach (i.e., two or more gene therapy strategies may be needed or gene therapy may be used in conjunction with reduced levels of conventional immunosuppression) (Wood and Fry, 1999).

6.4. Recurrent autoimmunity as a mechanism of β cell allograft failure

Autoimmune mechanisms are blamed for the destruction of pancreatic β cells in overt type 1 diabetes. The exact means, however, by which this selective destruction occurs in humans is unknown (Atkinson and Maclaren, 1994). In order to gain an insight into the human disease, various animal models have been developed and extensively studied; the often investigated one being the non-obese diabetic (NOD) mouse (Atkinson and Leiter, 1999).

It is clear that in the NOD mouse model, both $CD4^+$ and $CD8^+$ T cells are involved in β cell destruction. Athymic NOD mice and NOD.*scid* mice that lack these cells do not develop insulitis or diabetes, and treatment of NOD mice with anti-CD3 antibodies inhibits the development of disease (Miller et al., 1988; Dardenne et al., 1989; Christianson et al., 1993). Adoptive transfer studies demonstrated that splenic T cells from NOD mice transplanted into young or irradiated NOD mice, transfer diabetes and that the process requires both $CD4^+$ and $CD8^+$ T cells (Thivolet et al., 1991; Yagi et al., 1992; Katz et al., 1993). However, the specific role(s) for these cells in diabetogenesis remain(s) unclear. $CD4^+$ T cells are required for both the activation of $CD8^+$ T cells and for their recruitment into the pancreatic islets (Nagata et al., 1994; Verdaguer et al., 1996). $CD8^+$ T cells clearly play an important role as effectors in β cell destruction in vivo (Di Lorenzo et al., 1998). The mechanisms through which diabetogenic $CD8^+$ T cells impart an initial β cell insult is thought to occur through one or even two different lytic pathways (Chervonsky et al., 1997; Itoh et al., 1997; Kagi et al., 1997; Kurrer et al., 1997). In the Fas pathway, a T cell membrane ligand (i.e., FasL) upregulated during T cell receptor engagement, binds a target cell surface receptor (i.e., Fas) resulting in cellular apoptosis. Since Fas-deficient NOD mice fail to develop diabetes or insulitis, this pathway has been suggested as a major mechanism for cellular damage (Chervonsky et al., 1997; Itoh et al., 1997; Kurrer et al., 1997). In a second pathway, the direct effects of perforin and granzymes on the target cell cause cellular death. The perforin-deficient NOD mice develop insulitis but not diabetes suggesting that this pathway also can constitute a potential mechanism of β cell destruction (Kagi et al., 1997).

The types of cytokines produced by T cells also appear to play an important role in the pathogenesis of type 1 diabetes in NOD mice. The prevailing dogma is that T helper (Th)1 cytokines (e.g., IL-2, interferon [IFN]-γ) impart the development of diabetes whereas Th2 or Th3 cytokines (e.g., IL-4, IL-10, TGF-β) prevent the disease. However, the role of cytokines in the pathogenesis of this disorder is not straightforward. For example, treatment of NOD mice with anti-IFN-γ can prevent the development of diabetes whereas the genetic absence (i.e., knockout) of IFN-γ in NOD mice results in a delay of disease development but does not prevent it (Hultgren et al., 1996; Sarvetnick et al., 1988). Systemic administration of IL-4 or IL-10 and the transgenic expression of IL-4 on β cells prevent diabetes in NOD mice (Pennline et al., 1994; Mueller et al., 1996; Cameron et al., 1997). On the other hand, local expression of IL-10 in islets accelerates the development of diabetes in NOD mice and IL-4 knockout NOD mice do not show changes in the timing of disease onset (Wang et al., 1998). The interactions between cytokines appear complicated and diabetes development may depend on a fine balance of immunoregulatory T cells and their products. This topic will be discussed in detail later in this chapter.

While most of the research to date has focused on T cells, macrophages and dendritic cells constitute a significant part of the inflammatory islet infiltrate during the early stage of insulitis in NOD mice (Walker et al., 1988; Voorbij et al., 1989; Amano and Yoon, 1990; Jansen et al., 1994). Indeed, this islet infiltrate precedes the influx of T cells, natural killer (NK) cells and B lymphocytes (Lee et al., 1988). Recent evidence supports the contention that the three populations of antigen presenting cells (APC; macrophages, dendritic cells, B-lymphocytes) appear to be required for the development of the effector T cells that destroy β cells (Tisch and McDevitt, 1996).

These cells not only assist the T cells in the destruction of β cells, but also produce factors that may be directly toxic to β cells. Such factors include macrophage-derived soluble mediators like oxygen free radicals (e.g., superoxide dismutase, catalase, etc.) and various cytokines (e.g., IL-1β, tumor necrosis factor (TNF)-α, and IFN-γ). These cytokines, released from activated macrophages, have been demonstrated in vitro to be toxic to β cells (Mandrup-Poulsen et al., 1987; Pukel et al., 1988; Appels et al., 1989; Lortz et al., 2000). The common final phase of macrophage-induced β cell toxicity seems to be mediated by superoxide anion and hydrogen peroxide. Since β cells exhibit very low free radical scavenging activity, they are exquisitely sensitive to the oxidative stress (Malaisse et al., 1982; Asayama et al., 1986). As with allograft rejection, the mechanisms of recurrent autoimmunity are very complex and gene therapy, the subject of this text, represents a promising avenue to overcome at least some of the immunological obstacles.

6.5. Gene transfer into islet cells

The ability to transfer genes into islets and perhaps the insulin producing β cells appears to represent an excellent opportunity to utilize gene therapy for improving the clinical efficacy of islet cell transplantation. Unlike the case of other organ transplants (e.g., heart, liver), pancreatic islets can be cultured ex vivo for a prolonged time offering a unique opportunity for gene therapeutic manipulation prior to transplantation. A variety of methods have been utilized to date for ex vivo gene transfer into islets including both non-viral (e.g., biolistic projectile, electroporation, lipofection, etc.) and viral (e.g., adenoviral, lentiviral, rAAV) technologies (Giannoukakis et al., 1999; Levine and Leibowitz, 1999; Kapturczak et al., 2001). The results of such investigations have been mixed, yet successful production of a series of recombinant transgenes applicable for islet cell transplantation has been observed, examples of which are indicated in Table 6.2. Recently, the efficacy and safety of rAAV-mediated transgene delivery into islet cells has been demonstrated (Flotte et al., 2001). The rAAV did not affect the metabolic β cell function and viral transduction afforded a robust and long-term transgene expression. Moreover, bicistronic rAAV vector incorporating a viral internal ribosome entry site (IRES), to allow for translation of two coupled cDNAs from a single mRNA transcript, has been recently utilized to successfully transduce pancreatic islet cells with two reporter proteins (Kapturczak et al., 2002). This opens up further possibilities for concomitant delivery of more than one gene encoding cytoprotective and/or immunoregulatory molecules and in addition may offer superior clinical utility for preservation of β cell functionin transplanted islets. Further improvement potential to rAAV-mediated islet gene transfer has been brought by recent work examining the transduction efficacy of various AAV serotypes and generation of receptor-targeted rAAV particles (Loiler et al., 2003). Out of examined AAV serotypes (1, 2, 3, 4, and 5), murine islets showed highest levels of transduction with AAV1, whereas serotype 5 resulted in highest transgene expression levels in human islets. The transduction was furthermore greatly enhanced with modified rAAV particles displaying a ligand (ApoE) that targets the low-density lipoprotein receptor (LDL-R), which is abundantly expressed both on murine and human islets (Loiler et al., 2003).

TABLE 6.2
Examples of candidate transgenes with potential utility in the prevention of islet transplant rejection (see text for discussion)

Process	Transgene	Result	References
T cell activation	CTLA4-Ig	Heart, liver, and pancreas: Prolonged allograft survival	Olthoff et al. (1998); Uchikoshi et al. (1999); Yangs et al. (1999)
Immunomodulation	IL-4	β cells in transgenic mice: Prevention of insulitis and autoimmune diabetes	Mueller et al. (1996)
		Transduced islets: Prevention of insulitis of syngeneic islet allografts in NOD-SCID mice	Gallichan et al. (1998)
		Transgenic expression in β cells: Induction of islet Ag-specific Th2 T cells inhibiting the diabetogenic T cells in pancreas	Gallichan et al. (1999)
		Systemic delivery via rAAV does not protect from autoimmune diabetes	Goudy et al. (2001)
	vIL-10	Heart: Prolongation of allograft survival	Qin et al. (1997)
		Transgenic islets: High levels of vIL-10 inhibit allogenic lymphocyte proliferation	Benhamou et al. (1996)
	IL-10	Transgenic expression in β cells: No protection from insulitis in NOD back-cross	Balasa and Sarvetnick (1996)
		Transgenic islets: No protection from allograft rejection	Deng et al. (1997)

(*continued*)

TABLE 6.2 (continued)

Process	Transgene	Result	References
		Systemic delivery via rAAV protects from autoimmune diabetes, and protects transplanted islets from autoimmune destruction	Goudy et al. (2001); Zhang et al. (2003)
	TGF-β and IL-10	Islets: Prevention of autoimmune diabetes and increased islet allograft and xenograft survival	Deng et al. (1997)
	p40 subunit of IL-12	Islets: Prolonged syngeneic graft survival in NOD mice ($\downarrow$ in IFN-γ and $\uparrow$ in TGF-β at the transplant site)	Yasuda et al. (1998)
Cell death, apoptosis, and oxidative damage	MnSOD	Insulinoma cells: Protection against IL-1β induced cell injury	Hohmeier et al. (1998)
		Liver: Protection from ischemia-reperfusion damage ($\downarrow$ NFκB and AP-1)	Zwacka et al. (1998)
	Thioredoxin	Transgenic β cell expression: Protection from autoimmune and streptozotocin-induced diabetes	Hotta et al. (1998)
	Catalase	Islets: Protection from oxidative damage	Benhamou et al. (1998)
		Insulinoma cells: Protection from NO and O$^-$ toxicity	Tiedge et al. (1999)
		Insulinoma cells: Protection against cytokine-mediated injury	Lortz et al. (2000)
	Glutathion peroxidase	Insulinoma cells: Protection from NO and O$^-$ toxicity	Tiedge et al. (1999)
		Insulinoma cells: Protection against cytokine-mediated injury	Lortz et al. (2000)

Gene	Effect	Reference
Cu-Zn SOD	Insulinoma cells: Protection from NO and O toxicity	Tiedge et al. (1999)
	Insulinoma cells: Protection against cytokine-mediated injury	Lortz et al. (2000)
hsp70	Islets: Protection from IL-1β-mediated injury	Margulis et al. (1991)
Heme oxygenase-1	Islets: Protection from IL-1β-mediated injury	Ye and Laychock (1998)
Soluble IL-1 receptor antagonist (IRAP)	Islets: Protection against IL-1β induced β cell injury	Giannoukakis et al. (1999)
Soluble TNF-α receptor	Transgenic β cell expression: Prevention from insulitis and IDDM	Hunger et al. (1996, 1997)
FasL (membrane bound)	Kidney and liver: Prolonged allograft survival	Fandrich et al. (1998); Swenson et al. (1998)
	Islets: Neutrophil-mediated destruction	Kang et al. (1997)
FasL (soluble)	Islets: Prolonged allograft survival	Gainer et al. (1998)
Bcl-2	Liver: Reduction of cytotoxicity and apoptosis during cold ischemia	Bilbo et al. (1999)
	β cells in vitro: Prevention of cytokine-induced apoptosis	Liu et al. (1996)
	Islets in vitro: Prevention of cytokine-induced apoptosis	Rabinovitch et al. (1999)
Bcl-x$_L$	Transgenic islets: Prevention of cytokine-induced apoptosis and prolonged islet graft survival	Fujita et al. (2000)
Anti-apoptotic gene A20	Islets: Inhibition of cytokine-mediated apoptosis ($\downarrow$ NFκB)	Grey et al. (1999)

6.6. Potential utility of rAAV-mediated gene therapy for islet transplantation and prevention of autoimmunity recurrence in type 1 diabetes

The possibility of ex vivo genetic modification of islets, a process that could render them resistant to immune-mediated destruction, has become increasingly intriguing as new and more dependable transgene delivery systems such as rAAV emerge. For reasons described previously, the rAAV-based vectors are of particular interest for the transduction of pancreatic islets. The predominant question is which one of gene or genes should be subject to therapeutic testing in rAAV-based systems?

Current concepts of interfering with the course of rejection address practically all levels of T cell function including antigen recognition and presentation, activation, costimulation, immunoregulation, tolerance formation, and processes of cell death. Hence, gene therapy brings, at least on a conceptual level, virtually unlimited possibilities. In fact, research efforts to date have followed this possibility with an extensive number of molecules being subjected to various trials for therapeutic gene delivery for the prevention and reversal of type 1 diabetes. Table 6.2 illustrates approaches and transgenes believed to be potentially most useful, either alone or in combination, for ex vivo gene delivery into islets.

6.6.1. Interference with T cell activation

The so called "signal one"—a specific interaction between T cell receptor and the MHC-peptide complex—is one of the earliest steps common for both the initiation of allograft rejection and the T-cell mediated autoimmune reaction. An interaction between CD28 T cell surface molecules and B7 molecules (B7-1 or CD80; B7-2 or CD86) located on the membrane of APC ("signal two") is also required for a full activation of T cells. An alternative ligand for B7 molecules is the cytotoxic T lymphocyte antigen 4

(CTLA-4). Blocking of this interaction with a CTLA-4 fusion protein (CTLA-4Ig) has been shown to significantly prolong the survival of heart (Yangs et al., 1999), liver (Olthoff et al., 1998), and pancreas (Uchikoshi et al., 1999) allografts. In the field of islet transplantation, CTLA-4Ig transgene has been used to transfect syngeneic and allogeneic myoblasts, which were co-transplanted with allogeneic islets to provide local secretion of CTLA-4Ig (Chahine et al., 1995). Only syngeneic myoblasts provided enough local production of this transgene to result in prolonged islet allograft survival, whereas the allogeneic myoblasts were quickly destroyed without a chance to produce sufficient amounts of CTLA-4Ig. This destruction could have possibly been prevented with a transient pharmacological immunosuppression allowing for CTLA-4Ig synthesis and secretion to begin.

6.6.2. Immunomodulation

The concept of immunoregulation of cellular immune responses has gained significant research attention over the last decade. While autoimmune β cell destruction, at least in NOD mice, appears to be mediated by T cells (Bendelac et al., 1987; Miller et al., 1988), their development and activation appears to be due, in large part, to an intrinsic inability to induce various immunotolerogenic functions (Miller et al., 1988; Oldstone, 1988; Sadelain et al., 1990; Shehadeh et al., 1994). In NOD mice, the autoimmune tissue destruction appears to be promoted by a Th1 pattern of cytokine release from self-peptide reactive CD4$^+$ T cells. The Th1 cytokines, including IL-2 and IFN-γ, support macrophage activation, hypersensitivity responses, and immunoglobulin (Ig) isotype switching to IgG2a. In contrast, autoimmune tissue destruction seems to be suppressed when the self-reactive CD4$^+$ T cells produce a Th2 pattern of cytokines (e.g., IL-4, IL-5, IL-6, IL-10, and IL-13), which provide help for the activation of B lymphocytes-mediated humoral immunity and Ig isotype switching to IgG1 and IgE. Of the aforementioned cytokines, IL-4 appears to be the most important in switching CD4$^+$ T cells from Th1 to Th2 response profile (McArthur and Raulet, 1993; Parishes et al., 1993; Taylor-Robinson and Phillips, 1994). However, IL-10 also serves an important role by decreasing Th1, NK T cell, and macrophage functions as well as increasing B lymphocyte and macrophage activities.

Relatively late in the course of pre-diabetes in NOD mice a switch from Th2 to Th1 subsets appears to occur; converting the non-destructive lymphocytic infiltration of predominately Th2 activity into an aggressive destructive and pathogenic Th1 response (Rabinovitch, 1994; Bach, 1995; Liblau et al., 1995; Kroemer et al., 1996). Studies utilizing systemic administration of IL-4 and IL-10 have demonstrated that cytokines can be used to induce an immune deviation towards the Th2 phenotype and prevent diabetes in NOD mice (Rapoport et al., 1993; Pennline et al., 1994). These studies complement those indicating that detection of IL-4 in islets at the onset of inflammation identifies non-destructive insulitis (Arreaza et al., 1997), and that NOD mice with pancreatic (insulin promoter) expression of IL-4 are protected from autoimmunity (Mueller et al., 1996). Moreover, transgenic islet cell expression of IL-4 induces islet antigen specific Th2 cells that block the action of β cell reactive T cells in the pancreas (Gallichan et al., 1999), and may correct the aforementioned inherited defect in NOD mice of forming Th2 responses (Cameron et al., 1997). Although unproven, "the picture" for Th2 immunity in spontaneous development of type 1 diabetes seems promising. The concept when applied to islet cell transplantation is far less clear. The dominant host response to an islet allograft does appears to occur through Th1-like mechanisms, and Th-2 dominant patterns (e.g., diminished IFNγ/increased IL-4 and diminished IL-12/increased IL-10) are observed in long-term surviving grafts (Nickerson et al., 1994). However, a number of attempts to recreate the benefits of such a pattern through targeted islet cell cytokine expression have failed including IL-10 transgenic allografts (Wogensen et al., 1994), IL-4, and IL-10 (adenovirus) syngeneic grafts (Smith et al., 1997), and IL-4 transgenic allografts in vitro (Davies et al., 1999). Moreover, as shown in cases of xenogeneic transplantation, local IL-10 expression can even lead to accelerated graft rejection (Deng et al., 1997). In contrast, other reports argue for therapeutic effectiveness including systemic-therapy with IL-4 and IL-10 inhibiting diabetes recurrence in NOD mice transplanted with syngeneic islets (Rabinovitch et al., 1995), IL-4 transgenic islets resistant to disease when challenged with diabetogenic splenocytes (Mueller et al., 1996), and decreased alloreactivity in vitro to human islets secreting IL-10 (Benhamou et al., 1996). Perhaps most promising is the recent report demonstrating that syngeneic islet grafts expressing lentiviral mediated IL-4 are protected from insulitis in an adoptive transfer model (Gallichan et al., 1998). While sometimes at conflict, this collective body would suggest that the site of therapeutic administration, cytokine action, and concentration each play an important role in the success of therapeutic outcomes and may provide an explanation for the reportedly paradoxical effects. This concept finds support in other autoimmune models where local IL-10 administration reduces endotoxin-induced ocular inflammation whereas systemic delivery exacerbates disease pathology, and TGF-β provided locally induces arthritis yet systemic administration can attenuate inflammation [reviewed in Balasa and Sarvetnick (1996)]. In our own studies, we were able to demonstrate that rAAV-mediated skeletal muscle transduction of female NOD mice with IL-10 led to complete prevention of diabetes development (Goudy et al., 2001). This was not seen in respective IL-4-transduced mice. The rAAV-IL-10 transduction resulted in attenuation of insulin autoantibody production, quantitatively reduced insulitis, and maintained islet insulin content. Also, rAAV-mediated systemic gene therapy was successfully utilized to significantly inhibit the recurrence of type 1 diabetes in diabetic NOD mice treated with syngeneic islet transplantation (Zhang et al., 2003).

The site as well as timing of therapeutic exposure to immunomodulatory cytokines can present an important issue. The cytokine gene therapy strategies predominantly but not exclusively involve three modes of delivery (Schmidt-Wolf and Schmidt-Wolf, 1995; Robbins and Evans, 1996; Giannoukakis et al., 1999; Kapturczak et al., 2001). Cells targeted for autoimmune attack may be genetically modified to express cytokines that protect them from immune-mediated destruction (i.e., target tissue gene therapy). Another strategy allows ex vivo genetic alteration of autoreactive T cells to deliver anti-inflammatory cytokines to the sites of autoimmune processes (i.e., T-cell mediated gene therapy). Finally, new advances in muscle delivery offer the hope of efficient systemic cytokine production.

6.6.3. Prevention of oxidative injury

Although the induction of tolerance to an allograft or xenograft would be the ideal goal, a successful interference with the process of immune-mediated cellular death seems to be within the potential reach of currently available transgene delivery. Although not as elegant in conceptual sense, this approach would provide a sort of "generic" protection from a variety of cytotoxic processes.

One potential mechanism of β cell injury is thought to result from infiltration of T cells and macrophages involving direct exposure to reactive oxygen species (ROS) produced by those cells, especially peroxynitrite and superoxide. Moreover, the infiltrating cells secrete pro-inflammatory cytokines (e.g., IL1β, IFN-γ, TNF-α), which are thought to be involved in induction of cellular injury (apoptosis and necrosis) through increasing the production of ROS in β cells. This ROS-mediated injury is accentuated by the relatively low activities of critically relevant, cytoprotective, antioxidant enzymes in pancreatic islets, specifically manganese superoxide dismutase (MnSOD) and glutathion peroxidase (Lenzen et al., 1996; Tiedge et al., 1997).

As shown in Table 6.2, significant cytoprotective effects in β cells against cytokine- and/or ROS-mediated β cell injury have been demonstrated by overexpression of antioxidant enzymes like MnSOD, thioredoxin, catalase, glutathion peroxidase, and Cu-ZnSOD (Benhamou et al., 1998; Hohmeier et al., 1998; Hotta et al., 1998; Zwacka et al., 1998; Tiedge et al., 1999; Lortz et al., 2000). Also, upregulation of heat shock proteins (e.g., hsp32, hsp70) or HO–1 provided resistance against deleterious effects of IL–1β (Margulis et al., 1991; Ye and Laychock, 1998).

6.6.4. Interference with apoptotic cascade

While increasing the antioxidant potential of β cells appears to prevent cell death, the overexpression of proteins imparting islet cell resistance to immune based destruction through interference with the apoptotic pathway, also appears to be quite promising. Increased levels of Bcl-2 (Liu et al., 1996; Bilbao et al., 1999; Rabinovitch et al., 1999), Bcl-x_L (Fujita et al., 2000) and anti-apoptotic protein A20 (Grey et al., 1999) were all shown to protect β cells from apoptosis.

Another potential way of preventing the cell death process is to interfere with cytokine effects at the receptor level. Examples of such interference include the β cell overexpression of soluble antagonists of cytokine receptors as has been elegantly demonstrated for both IL-1β (Giannoukakis et al., 1999) and TNF-α (Hunger et al., 1996, 1997) receptors. A similar approach was applied in an attempt to interfere with Fas-mediated β cell apoptosis wherein the pancreatic islet secretion of soluble Fas ligand was shown to prolong the survival of islet allografts (Gainer et al., 1998). Interestingly, membrane bound Fas ligand expressed in islets not only fails to protect them from immune destruction but even accelerates their destruction in a neutrophil-mediated process (Kang et al., 1997). This is not true for other solid organ transplants as demonstrated with kidney (Swenson et al., 1998) and liver (Fandrich et al., 1998) allografts.

6.7. Progress in insulin replacement strategies utilizing gene therapy

As previously noted, islet cell transplantation can be highly successful resulting in proper glycemic control in the absence of exogenous insulin (Shapiro et al., 2000; Kojima et al., 2003). The lack of sufficient number of islets required to treat more than a million type 1 diabetes patients is a major obstacle. This has resulted in numerous strategies to develop an alternative method of insulin replacement either through gene therapy or islet cell neogenesis. Two dominant alternative strategies for insulin replacement therapy, not including islet cell transplantation, can be classified as either: (1) utilizing gene therapy to impart insulin production within a surrogate organ or tissue, or (2) cellular conversion of either pancreatic, hepatic, or embryonic stem cells to an "islet-like" population capable of insulin production. The following section describes both insulin replacement strategies as a potential treatment for type 1 and/or type 2 diabetes.

6.7.1. Hepatic insulin gene therapy

Gene therapy is defined as the transgenic delivery of genetic material via, in most cases, a non-pathogenic viral vector for treatment, cure, or prevention of a disease or clinical condition. Type 1 diabetes represents an excellent candidate for treatment with gene therapy since it is caused by the lack of a single protein (insulin). Multiple investigators have demonstrated successful insulin expression in cells other than islets, both in vivo and in vitro, in rat and mouse hepatocytes, K-cells, muscle, and various cell lines (Kaneda et al., 1989; Valera et al., 1994; Kolodka et al., 1995; Short et al., 1998; Tuch et al., 1998; Falqui et al., 1999; Gros et al., 1999; Chen et al., 2000; Cheung et al., 2000; Lee et al., 2000; Shapiro et al., 2000; Thule and Liu, 2000; Thule et al., 2000; Auricchio et al., 2002; Shaw et al., 2002). Transgenic insulin gene delivery has been achieved through a variety of viral vectors which include rAAV

(Lee et al., 2000), adenovirus (Chen et al., 2000; Thule and Liu, 2000), and retrovirus (Kolodka et al., 1995). In addition, insulin secretion has also been achieved via plasmid-mediated gene transfer into rat skeletal muscle in vivo (Shaw et al., 2002). Unlike pancreatic β cells, most other cells (exception being K-cells) lack the prohormone convertases PC2 and PC3 necessary for enzymatic processing of insulin. Therefore, modifications have been made to the insulin gene to allow for transgenic expression in other non-islet cell types. This includes modifying the PC2 and PC3 enzymatic sites of preproinsulin to be recognized and cleaved by furin (endogenous protease found in Golgi apparatus) that results in successful biological processing of cells lacking PC2 and PC3 (Groskreutz et al., 1994). Another strategy involves the use of an insulin single chain analog that does not require biological processing, however has significantly reduced biological activity (Lee et al., 2000).

Transgenic insulin expression must be directly concordant with blood glucose concentration to be ultimately successful. Hence, the liver has been targeted as a surrogate organ for insulin expression via gene therapy. The liver is highly glucose responsive due to the expression of glucose transporter-2 (GLUT-2) and glucokinase (Burcelin et al., 2000; Iynedjian, 1993). In addition, the liver is directly involved in glucose regulation and metabolism, and therefore multiple liver specific promoters are glucose responsive, particularly those involved with the gluconeogenic pathway (e.g., glucose 6-phosphatase [G6P] and L-pyruvate kinase [L-PK]). Several investigators have engineered glucose responsive constructs typically containing a gluconeogenic promoter controlling the regulation of the furin cleavable human preproinsulin gene (fHPI) or the single chain insulin analogue (SIA). Some examples have included G6P-fHPI and L-PK-SIA delivered by either adenoviral or adeno-associated viral vectors (Chen et al., 2000, 2001; Yoon and Jun, 2002). These constructs are capable of glucose-stimulated and insulin-inhibitory insulin expression in liver-derived cells lines (i.e., H4IIE rat hepatoma cells), rat hepatocytes, or animal models. In addition, these constructs are capable of dramatically reducing hyperglycemia in diabetic rats and mice, and in some cases establishing complete normoglycemia delivered by adenovirus or rAAV. However, the most dramatic deficiency of hepatic insulin production is the inability to mimic the rapid and robust kinetics of glucose-responsive insulin release of islet cells. This has resulted in suboptimal glucose responsive insulin release via hepatic insulin gene therapy. Therefore, there is still the need to improve the kinetics of hepatic insulin production to treat type 1 diabetes and there are currently other approaches utilizing gene therapy to induce the liver or even other cell types to impart insulin production.

6.7.2. Islet cell neogenesis

The liver and the pancreas share many similar characteristics and even arise from adjacent areas in the endoderm of the developing embryo (Wells and Melton, 1999). It is believed that the pancreas and liver both arise from the ventral foregut endoderm, and the default pathway during embryogenesis is towards pancreatic development. However, fibroblast growth factor-like signal released by the cardiac mesoderm transfers the development of certain cells into hepatocytes. This suggests that these two cell types arise from a bipotent stem cell and differ from one another by one key developmental step. Therefore, transdifferentiation between the liver and pancreas is highly plausible and actively investigated as a potential method of islet cell neogenesis. Several transcriptional factors have been identified as potent activators for the development of the pancreas and subsequent differentiation from the liver. These transcriptional factors include pancreatic duodenal homeobox-1 (*Pdx-1*; also known as *Ipf1*), and NeuroD (*NeuroD*, also known as *BETA2*) among many others (Naya et al., 1997; Kim and Hebrok, 2001; McKinnon and Docherty, 2001). This has led investigators to employ gene therapy to deliver these islet cell differentiating factors into animal models to essentially reprogram the liver or other tissue into insulin producing "islet-like" cells.

Ferber et al. (2000) have reported utilizing recombinant adenovirus to deliver the transgene of Pdx-1 to the livers of BALB/C and C57BL6 mice to reprogram hepatocytes into cells that share β cell characteristics including insulin production and expression of islet cell specific genes. Recombinant adenovirus-mediated gene transfer of PDX-1 resulted in expression of the genes for mouse insulin 1 and 2, and prohormone convertases of 1/3 (PC 1/3). In addition, the liver successfully produced biologically active insulin and ameliorated hyperglycemia in STZ-treated diabetic mice. However, despite a dramatic decrease in blood glucose levels of the diabetic mice (from 600 to 200 mg/dl after 1 week of receiving adenovirus delivered PDX-1) there were no instances of established normoglycemia during the course of the experiment. Kojima et al. (2003) evaluated the use of helper-dependant adenovirus to deliver both NeuroD alone and in conjunction with a β cell stimulating hormone known as betacellulin (*Btc*) to STZ-treated mice. Diabetes was partially but not completely reversed in STZ-treated mice administered with NeuroD alone. However, STZ-treated mice given both helper-dependant administered NeuroD and Btc achieved complete reversal of diabetes and were normoglycemic for the duration of the experiment (i.e., >120 days). In addition, the presence of insulin and other β cell specific transcripts including the proinsulin-processing enzymes, and glucokinase were detected in the liver. This *NeuroD-Btc* gene therapy regimen demonstrated successful induction of islet neogenesis and a potential novel combination therapy as a treatment for type 1 diabetes. The intricacies of both insulin secretion and the embryonic tissue differentiation of the pancreatic β cell are highly complex and strictly regulated. A suitable treatment for type 1 diabetes that would entail mimicking insulin secretion (hepatic insulin gene therapy) or transdifferentiating β cells (islet cell neogenesis) from other tissues (i.e., liver), may require strategies that are as complicated and synchronized as the systems investigators are attempting to imitate and potentially replace. Future gene therapy strategies to treat type 1 diabetes may be comprised of "shotgun" approaches that entail transgenic delivery of multiple β cell specific transcriptional factors delivered with vectors (i.e., rAAV) that are highly efficient at transduction of specific target tissue and above all safe for patients.

6.8. Summary and future directions

The case for developing novel methods of gene therapy for the prevention of recurrent diabetes following islet cell transplantation is well established given the extensive and unfortunate history of clinical failures as well as the side effects associated with chronic immunosuppression. Gene therapy approaches aimed at the creation of insulin producing apparatus in surrogate tissues and at islet cell neogenesis from stem cells hold a great promise in circumventing some of the immunological problems associated with islet cell transplantation. Many vectors are currently being subjected to investigation in an attempt to improve gene delivery into islets and ensure patient safety. Viral vectors have shown the ability to target and enter host cells with a high degree of efficiency, and therefore form the subject of most such efforts, with recent improvements in the propagation and delivery of rAAV, causing major enthusiasm. At present, many questions exist that must first be addressed in animal models (Table 6.3). Provided that

TABLE 6.3
Outstanding questions for gene therapy in islet cell transplantation

- Which gene or genes will be most effective in preventing and/or delaying islet cell transplant rejection?
- Will genes for allograft rejection prevent recurrent autoimmunity and vice versa?
- How safe will islet immunomodulatory therapy be in terms of overall physiology?
- Would such immunomodulatory therapies totally avoid the need for conventional immunosuppression?
- What level of transgene expression and what percentage of cells require transfection in order to provide protection from disease?
- Can surrogate tissues engineered to produce insulin avoid immune-mediated damage?

sufficient answers to these and other questions are found, gene therapy can be extended to patients with existing islet cell transplantation programs in conjunction with conventional immunosuppression. The future of gene therapy for type 1 diabetes poses an exciting challenge and one that raises important questions regarding the retention of graft function in an environment of allograft rejection as well as recurrent autoimmunity.

References

Alejandro, R., Lehmann, R., Ricordi, C., Kenyon, N. S., Angelico, M. C., Burke, G., Esquenazi, V., Nery, J., Betancourt, A. E., Kong, S. S., Miller, J. and Mintz, D. H. (1997). Long-term function (6 years) of islet allografts in type 1 diabetes. Diabetes 46, 1983–1989.

Amano, K. and Yoon, J. W. (1990). Studies on autoimmunity for initiation of beta-cell destruction. V. Decrease of macrophage-dependent T lymphocytes and natural killer cytotoxicity in silica-treated BB rats.Diabetes 39, 590–596.

Appels, B., Burkart, V., Kantwerk-Funke, G., Funda, J., Kolb-Bachofen, V. and Kolb, H. (1989). Spontaneous cytotoxicity of macrophages against pancreatic islet cells. J. Immunol. 142, 3803–3808.

Arreaza, G. A., Cameron, M. J., Jaramillo, A., Gill, B. M., Hardy, D., Laupland, K. B., Rapoport, M. J., Zucker, P., Chakrabarti, S., Chensue, S. W., Qin, H. Y., Singh, B. and Delovitch, T. L. (1997). Neonatal activation of CD28 signaling overcomes T cell anergy and prevents autoimmune diabetes by an IL-4-dependent mechanism. J. Clin. Invest. 100, 2243–2253.

Asayama, K., Kooy, N. W. and Burr, I. M. (1986). Effect of Vitamin E deficiency and selenium deficiency on insulin secretory reserve and free radical scavenging systems in islets: Decrease of islet manganosuperoxide dismutase. J. Lab. Clin. Med. 107, 459–464.

Atkinson, M. A. and Leiter, E. H. (1999). The NOD mouse model of type 1 diabetes: As good as it gets? Nat. Med. 5, 601–604.

Atkinson, M. A. and Maclaren, N. K. (1994). The pathogenesis of insulin-dependent diabetes mellitus. N. Engl. J. Med. 331, 1428–1436.

Auricchio, A., Gao, G. P., Yu, Q. C., Raper, S., Rivera, V. M., Clackson, T. and Wilson, J. M. (2002). Constitutive and regulated expression of processed insulin following in vivo hepatic gene transfer. Gene Ther. 9, 963–971.

Bach, J. F. (1995). Insulin-dependent diabetes mellitus as a beta-cell targeted disease of immunoregulation. J. Autoimmun. 8, 439–463.

Balasa, B. and Sarvetnick, N. (1996). The paradoxical effects of interleukin 10 in the immunoregulation of autoimmune diabetes. J. Autoimm. 9, 283–286.

Bendelac, A., Carnaud, C., Boitard, C. and Bach, J. F. (1987). Syngeneic transfer of autoimmune diabetes from diabetic NOD mice to healthy neonates. Requirement for both L3T4+ and Lyt-2+ T cells.J. Exp. Med. 166, 823–832.

Benhamou, P. Y., Moriscot, C., Richard, M. J., Kerr-Conte, J., Pattou, F., Chroboczek, J., Lemarchand, P. and Halimi, S. (1998). Adenoviral-mediated catalase gene transfer protects porcine and human islets in vitro against oxidative stress. Transplant Proc. 30, 459.

Benhamou, P. Y., Mullen, Y., Shaked, A., Bahmiller, D. and Csete, M. E. (1996). Decreased alloreactivity to human islets secreting recombinant viral interleukin 10. Transplantation 62, 1306–1312.

Bilbao, G., Contreras, J. L., Gomez-Navarro, J., Eckhoff, D. E., Mikheeva, G., Krasnykh, V., Hynes, T., Thomas, F. T., Thomas, J. M. and Curiel, D. T. (1999). Genetic modification of liver grafts with an adenoviral vector encoding the Bcl-2 gene improves organ preservation. Transplantation 67, 775–783.

Bottino, R., Fernandez, L. A., Ricordi, C., Lehmann, R., Tsan, M. F., Oliver, R. and Inverardi, L. (1998). Transplantation of allogeneic islets of Langer-hans in the rat liver: Effects of macrophage depletion on graft survival and microenvironment activation. Diabetes 47, 316–323.

Burcelin, R., Dolci, W. and Thorens, B. (2000). Glucose sensing by the hepatoportal sensor is GLUT2-dependent: In vivo analysis in GLUT2-null mice. Diabetes 49, 1643–1648.

Cameron, M. J., Arreaza, G. A., Zucker, P., Chensue, S. W., Strieter, R. M., Chakrabarti, S. and Delovitch, T. L. (1997). IL-4 prevents insulitis and insulin-dependent diabetes mellitus in nonobese diabetic mice by potentiation of regulatory T helper–2 cell function. J. Immunol. 159, 4686–4692.

Chahine, A. A., Yu, M., McKernan, M. M., Stoeckert, C. and Lau, H. T. (1995). Immunomodulation of pancreatic islet allografts in mice with CTLA4Ig secreting muscle cells. Transplantation 59, 1313–1318.

Chen, R., Meseck, M. L. and Woo, S. L. (2001). Auto-regulated hepatic insulin gene expression in type 1 diabetic rats. Mol. Ther. 3, 584–590.

Chen, R., Meseck, M., McEvoy, R. C. and Woo, S. L. (2000). Glucose-stimulated and self-limiting insulin production by glucose 6-phosphatase promoter driven insulin expression in hepatoma cells. Gene Ther. 7, 1802–1809.

Chervonsky, A. V., Wang, Y., Wong, F. S., Visintin, I., Flavell, R. A., Janeway, C. A. J. and Matis, L. A. (1997). The role of Fas in autoimmune diabetes. Cell 89, 17–24.
Cheung, A. T., Dayanandan, B., Lewis, J. T., Korbutt, G. S., Rajotte, R. V., Bryer-Ash, M., Boylan, M. O., Wolfe, M. M. and Kieffer, T. J. (2000). Glucose-dependent insulin release from genetically engineered K cells. Science 290, 1959–1962.
Christianson, S. W., Shultz, L. D. and Leiter, E. H. (1993). Adoptive transfer of diabetes into immunodeficient NOD-scid/scid mice. Relative contributions of CD4+ and CD8+ T-cells from diabetic versus prediabetic NOD. NON-Thy-1a donors.Diabetes 42, 44–55.
Dardenne, M., Lepault, F., Bendelac, A. and Bach, J. F. (1989). Acceleration of the onset of diabetes in NOD mice by thymectomy at weaning. Eur. J. Immunol. 19, 889–895.
Davies, J. D., Mueller, R., Minson, S., O'Connor, E., Krahl, T. and Sarvetnick, N. (1999). Interleukin-4 secretion by the allograft fails to affect the allograft-specific interleukin-4 response in vitro. Transplantation 67, 1583–1589.
Deng, S., Ketchum, R. J., Yang, Z. D., Kucher, T., Weber, M., Shaked, A., Naji, A. and Brayman, K. L. (1997). IL-10 and TGF-beta gene transfer to rodent islets: Effect on xenogeneic islet graft survival in naive and B-cell-deficient mice. Transplant Proc. 29, 2207–2208.
Deng, S., Yang, Z. D., Ketchum, R. J., Kucher, T., Weber, M., Shaked, A., Naji, A. and Brayman, K. L. (1997). Transfer of genes for IL-10 and TGF-beta to isolated human pancreatic islets. Transplant Proc. 29, 2206.
Di Lorenzo, T. P., Graser, R. T., Ono, T., Christianson, G. J., Chapman, H. D., Roopenian, D. C., Nathenson, S. G. and Serreze, D. V. (1998). Major histocompatibility complex class I-restricted T cells are required for all but the end stages of diabetes development in nonobese diabetic mice and use a prevalent T cell receptor alpha chain gene rearrangement. Proc. Natl. Acad. Sci. USA 95, 12538–12543.
Dunn, D. L. (1990). Problems related to immunosuppression. Infection and malignancy occurring after solid organ transplantation.Crit. Care Clin. 6, 955–977.
Falqui, L., Martinenghi, S., Severini, G. M., Corbella, P., Taglietti, M. V., Arcelloni, C., Sarugeri, E., Monti, L. D., Paroni, R., Dozio, N., Pozza, G. and Bordignon, C. (1999). Reversal of diabetes in mice by implantation of human fibroblasts genetically engineered to release mature human insulin. Hum. Gene Ther. 10, 1753–1762.
Fandrich, F., Lin, X., Zhu, X., Kloppel, G., Parwaresch, R. and Kremer, B. (1998). CD95L confers immune privilege to liver grafts which are spontaneously accepted. Transplant Proc. 30, 1057–1058.

Ferber, S., Halkin, A., Cohen, H., Ber, I., Einav, Y., Goldberg, I., Barshack, I., Seijffers, R., Kopolovic, J., Kaiser, N. and Karasik, A. (2000). Pancreatic and duodenal homeobox gene 1 induces expression of insulin genes in liver and ameliorates streptozotocin-induced hyperglycemia. Nat. Med. 6, 568–572.

Flotte, T., Agarwal, A., Wang, J., Song, S., Fenjves, E. S., Inverardi, L., Chesnut, K., Afione, S., Loiler, S., Wasserfall, C., Kapturczak, M., Ellis, T., Nick, H. and Atkinson, M. (2001). Efficient ex vivo transduction of pancreatic islet cells with recombinant adeno-associated virus vectors. Diabetes 50, 515–520.

Fujita, J., Zhou, J.-P., Szot, G., Ostrega, D., Baldwin, A., Park, C.-G., Thompson, C., Bluestone, J. and Polonsky, K. (2000). Bcl-x_L overexpression prevents cytokine-induced dysfunction and apoptosis in pancreatic b cells and prolongs graft survival in islet transplantation. Diabetes 49 (Abstract).

Gainer, A. L., Suarez-Pinzon, W. L., Min, W. P., Swiston, J. R., Hancock-Friesen, C., Korbutt, G. S., Rajotte, R. V., Warnock, G. L. and Elliott, J. F. (1998). Improved survival of biolistically transfected mouse islet allografts expressing CTLA4-Ig or soluble Fas ligand. Transplantation 66, 194–199.

Gallichan, W. S., Balasa, B., Davies, J. D. and Sarvetnick, N. (1999). Pancreatic IL-4 expression results in islet-reactive Th2 cells that inhibit diabetogenic lymphocytes in the nonobese diabetic mouse. J. Immunol. 163, 1696–1703.

Gallichan, W. S., Kafri, T., Krahl, T., Verma, I. M. and Sarvetnick, N. (1998). Lentivirus-mediated transduction of islet grafts with interleukin 4 results in sustained gene expression and protection from insulitis. Hum. Gene Ther. 9, 2717–2726.

Giannoukakis, N., Rudert, W. A., Ghivizzani, S. C., Gambotto, A., Ricordi, C., Trucco, M. and Robbins, P. D. (1999). Adenoviral gene transfer of the interleukin-1 receptor antagonist protein to human islets prevents IL-1beta-induced beta-cell impairment and activation of islet cell apoptosis in vitro. Diabetes 48, 1730–1736.

Giannoukakis, N., Rudert, W. A., Robbins, P. D. and Trucco, M. (1999). Targeting autoimmune diabetes with gene therapy. Diabetes 48, 2107–2121.

Goudy, K., Song, S., Wasserfall, C., Zhang, Y. C., Kapturczak, M., Muir, A., Powers, M., Scott-Jorgensen, M., Campbell-Thompson, M., Crawford, J. M., Ellis, T. M., Flotte, T. R. and Atkinson, M. A. (2001). Adeno-associated virus vector-mediated IL-10 gene delivery prevents type 1 diabetes in NOD mice. Proc. Natl. Acad. Sci. USA 98, 13913–13918.

Grey, S. T., Arvelo, M. B., Hasenkamp, W., Bach, F. H. and Ferran, C. (1999). A20 inhibits cytokine-induced apoptosis and nuclear factor kappa B-dependent gene activation in islets. J. Exp. Med. *190*, 1135–1145.

Gros, L., Riu, E., Montoliu, L., Ontiveros, M., Lebrigand, L. and Bosch, F. (1999). Insulin production by engineered muscle cells. Hum. Gene Ther. *10*, 1207–1217.

Groskreutz, D. J., Sliwkowski, M. X. and Gorman, C. M. (1994). Genetically engineered proinsulin constitutively processed and secreted as mature, active insulin. J. Biol. Chem. *269*, 6241–6245.

Hahn, H. J., Laube, F., Lucke, S., Kloting, I., Kohnert, K. D. and Warzock, R. (1986). Toxic effects of cyclosporine on the endocrine pancreas of Wistar rats. Transplantation *41*, 44–47.

Hering, B. J., Browatzki, C. C., Schultz, A., Bretzel, R. G. and Federlin, K. F. (1993). Clinical islet transplantation—registry report, accomplishments in the past and future research needs. Cell Transplant. *2*, 269–282.

Hirano, Y., Fujihira, S., Ohara, K., Katsuki, S. and Noguchi, H. (1992). Morphological and functional changes of islets of Langerhans in FK506-treated rats. Transplantation *53*, 889–894.

Hohmeier, H. E., Thigpen, A., Tran, V. V., Davis, R. and Newgard, C. B. (1998). Stable expression of manganese superoxide dismutase (MnSOD) in insulinoma cells prevents IL-1 beta-induced cytotoxicity and reduces nitric oxide production. J. Clin. Invest. *101*, 1811–1820.

Hotta, M., Tashiro, F., Ikegami, H., Niwa, H., Ogihara, T., Yodoi, J. and Miyazaki, J. (1998). Pancreatic beta cell-specific expression of thioredoxin, an antioxidative and antiapoptotic protein, prevents autoimmune and Streptozotocin-induced diabetes. J. Exp. Med. *188*, 1445–1451.

Hultgren, B., Huang, X., Dybdal, N. and Stewart, T. A. (1996). Genetic absence of gamma-interferon delays but does not prevent diabetes in NOD mice. Diabetes *45*, 812–817.

Hunger, R. E., Carnaud, C., Garcia, I., Vassalli, P. and Mueller, C. (1997). Prevention of autoimmune diabetes mellitus in NOD mice by transgenic expression of soluble tumor necrosis factor receptor p55. Eur. J. Immunol. *27*, 255–261.

Hunger, R. E., Muller, S., Laissue, J. A., Hess, M. W., Carnaud, C., Garcia, I. and Mueller, C. (1996). Inhibition of submandibular and lacrimal gland infiltration in nonobese diabetic mice by transgenic expression of soluble TNF-receptor p55. J. Clin. Invest. *98*, 954–961.

Itoh, N., Imagawa, A., Hanafusa, T., Waguri, M., Yamamoto, K., Iwahashi, H., Moriwaki, M., Nakajima, H., Miyagawa, J., Namba, M., Makino, S., Nagata, S., Kono, N. and Matsuzawa, Y. (1997). Requirement of Fas for the development of autoimmune diabetes in nonobese diabetic mice. J. Exp. Med. *186*, 613–618.

Iynedjian, P. B. (1993). Mammalian glucokinase and its gene. Biochem. J. *293(Pt. 1)*, 1–13.

Jansen, A., Homo-Delarche, F., Hooijkaas, H., Leenen, P. J., Dardenne, M. and Drexhage, H. A. (1994). Immunohistochemical characterization of monocytes-macrophages and dendritic cells involved in the initiation of the insulitis and beta-cell destruction in NOD mice. Diabetes *43*, 667–675.

Jindal, R. M. (1994). Posttransplant diabetes mellitus—a review. Transplantation *58*, 1289–1298.

Kagi, D., Odermatt, B., Seiler, P., Zinkernagel, R. M., Mak, T. W. and Hengartner, H. (1997). Reduced incidence and delayed onset of diabetes in perforin-deficient nonobese diabetic mice. J. Exp. Med. *186*, 989–997.

Kaneda, Y., Iwai, K. and Uchida, T. (1989). Introduction and expression of the human insulin gene in adult rat liver. J. Biol. Chem. *264*, 12126–12129.

Kang, S. M., Schneider, D. B., Lin, Z. H., Hanahan, D., Dichek, D. A., Stock, P. G. and Baekkeskov, S. (1997). Fas ligand expression in islets of Langerhans does not confer immune privilege and instead targets them for rapid destruction. Nat. Med. *3*, 738–743.

Kapturczak, M. H., Flotte, T. and Atkinson, M. A. (2001). Adeno-associated virus (AAV) as a vehicle for therapeutic gene delivery: Improvements in vector design and viral production enhance potential to prolong graft survival in pancreatic islet cell transplantation for the reversal of type 1 diabetes. Curr. Mol. Med. *1*, 245–258.

Kapturczak, M., Zolotukhin, S., Cross, J., Pileggi, A., Molano, R. D., Jorgensen, M., Byrne, B., Flotte, T., Ellis, T. R., Inverardi, L., Ricordi, C., Nick, H., Atkinson, M. and Agarwal, A. (2002). Transduction of human and mouse pancreatic islet cells using a bicistronic recombinant adeno-associated viral vector. Mol. Ther. *5*, 154–160.

Katz, J., Benoist, C. and Mathis, D. (1993). Major histocompatibility complex class I molecules are required for the development of insulitis in non-obese diabetic mice. Eur. J. Immunol. *23*, 3358–3360.

Kaufman, D. B., Platt, J. L., Rabe, F. L., Dunn, D. L., Bach, F. H. and Sutherland, D. E. (1990). Differential roles of Mac-1+ cells, and CD4+ and CD8+ T lymphocytes in primary nonfunction and classic rejection of islet allografts. J. Exp. Med. *172*, 291–302.

Kenyon, N. S., Alejandro, R., Mintz, D. H. and Ricordi, C. (1996). Islet cell transplantation: Beyond the paradigms. Diabetes Metab. Rev. *12*, 361–372.

Kenyon, N. S., Chatzipetrou, M., Masetti, M., Ranuncoli, A., Oliveira, M., Wagner, J. L., Kirk, A. D., Harlan, D. M., Burkly, L. C. and Ricordi, C. (1999). Long-term survival and function of intrahepatic islet allografts in rhesus monkeys treated with humanized anti-CD154. Proc. Natl. Acad. Sci. USA 96, 8132–8137.

Kim, S. K. and Hebrok, M. (2001). Intercellular signals regulating pancreas development and function. Genes Dev. 15, 111–127.

Kojima, H., Fujimiya, M., Matsumura, K., Younan, P., Imaeda, H., Maeda, M. and Chan, L. (2003). NeuroD-betacellulin gene therapy induces islet neogenesis in the liver and reverses diabetes in mice. Nat. Med. 9, 596–603.

Kolodka, T. M., Finegold, M., Moss, L. and Woo, S. L. (1995). Gene therapy for diabetes mellitus in rats by hepatic expression of insulin. Proc. Natl. Acad. Sci. USA 92, 3293–3297.

Kroemer, G., Hirsch, F., Gonzalez-Garcia, A. and Martinez, C. (1996). Differential involvement of Th1 and Th2 cytokines in autoimmune diseases. Autoimmunity 24, 25–33.

Kurrer, M. O., Pakala, S. V., Hanson, H. L. and Katz, J. D. (1997). Beta cell apoptosis in T cell-mediated autoimmune diabetes. Proc. Natl. Acad. Sci. USA 94, 213–218.

Lee, H. C., Kim, S. J., Kim, K. S., Shin, H. C. and Yoon, J. W. (2000). Remission in models of type 1 diabetes by gene therapy using a single-chain insulin analogue. Nature 408, 483–488.

Lee, K. U., Amano, K. and Yoon, J. W. (1988). Evidence for initial involvement of macrophage in development of insulitis in NOD mice. Diabetes 37, 989–991.

Lenzen, S., Drinkgern, J. and Tiedge, M. (1996). Low antioxidant enzyme gene expression in pancreatic islets compared with various other mouse tissues. Free Radic. Biol. Med. 20, 463–466.

Levine, F. and Leibowitz, G. (1999). Towards gene therapy of diabetes mellitus. Mol. Med. Today 5, 165–171.

Liblau, R. S., Singer, S. M. and McDevitt, H. O. (1995). Th1 and Th2 CD4+ T cells in the pathogenesis of organ-specific autoimmune diseases. Immunol. Today 16, 34–38.

Liu, Y. X., Rabinovitch, A., SuarezPinzon, W., Muhkerjee, B., Brownlee, M., Edelstein, D. and Federoff, H. J. (1996). Expression of the bcl-2 gene from a defective HSV-1 amplicon vector protects pancreatic beta-cells from apoptosis. Hum. Gene Ther. 7, 1719–1726.

Loiler, S. A., Conlon, T. J., Song, S., Warrington, K. H., Agarwal, A., Kapturczak, M., Li, C., Ricordi, C., Atkinson, M. A., Muzyczka, N. and Flotte, T. R. (2003). Targeting recombinant adeno-associated virus vectors to enhance gene transfer to pancreatic islets and liver. Gene Ther. 10, 1551–1558.

Lortz, S., Tiedge, M., Nachtwey, T., Karlsen, A. E., Nerup, J. and Lenzen, S. (2000). Protection of insulin-producing RINm5F cells against cytokine-mediated toxicity through overexpression of antioxidant enzymes. Diabetes 49, 1123–1130.

Malaisse, W. J., Malaisse-Lagae, F., Sener, A. and Pipeleers, D. G. (1982). Determinants of the selective toxicity of alloxan to the pancreatic B cell. Proc. Natl. Acad. Sci. USA 79, 927–930.

Mandrup-Poulsen, T., Bendtzen, K., Dinarello, C. A. and Nerup, J. (1987). Human tumor necrosis factor potentiates human interleukin 1-mediated rat pancreatic beta-cell cytotoxicity. J. Immunol. 139, 4077–4082.

Margulis, B. A., Sandler, S., Eizirik, D. L., Welsh, N. and Welsh, M. (1991). Liposomal delivery of purified heat shock protein hsp70 into rat pancreatic islets as protection against interleukin 1 beta-induced impaired beta-cell function. Diabetes 40, 1418–1422.

Masetti, M., Inverardi, L., Ranuncoli, A., Iaria, G., Lupo, F., Vizzardelli, C., Kenyon, N. S., Alejandro, R. and Ricordi, C. (1997). Current indications and limits of pancreatic islet transplantation in diabetic nephropathy. J. Nephrol. 10, 245–252.

McArthur, J. G. and Raulet, D. H. (1993). CD28-induced costimulation of T helper type 2 cells mediated by induction of responsiveness to interleukin 4. J. Exp. Med. 178, 1645–1653.

McKinnon, C. M. and Docherty, K. (2001). Pancreatic duodenal homeobox-1, PDX–1, a major regulator of beta cell identity and function. Diabetologia 44, 1203–1214.

Miller, B. J., Appel, M. C., O' Neil, J. J. and Wicker, L. S. (1988). Both the Lyt-2+ and L3T4+ T cell subsets are required for the transfer of diabetes in nonobese diabetic mice. J. Immunol. 140, 52–58.

Mueller, R., Krahl, T. and Sarvetnick, N. (1996). Pancreatic expression of interleukin-4 abrogates insulitis and autoimmune diabetes in nonobese diabetic (NOD) mice. J. Exp. Med. 184, 1093–1099.

Nagata, M., Santamaria, P., Kawamura, T., Utsugi, T. and Yoon, J. W. (1994). Evidence for the role of CD8+ cytotoxic T cells in the destruction of pancreatic beta-cells in nonobese diabetic mice. J. Immunol. 152, 2042–2050.

Naya, F. J., Huang, H. P., Qiu, Y., Mutoh, H., DeMayo, F. J., Leiter, A. B. and Tsai, M. J. (1997). Diabetes, defective pancreatic morphogenesis, and abnormal enteroendocrine differentiation in BETA2/neuroD-deficient mice. Genes Dev. 11, 2323–2334.

Nickerson, P., Steurer, W., Steiger, J., Zheng, X., Steele, A. W. and Strom, T. B. (1994). Cytokines and the Th1/Th2 paradigm in transplantation. Curr. Opin. Immunol. 6, 757–764.

Oldstone, M. B. (1988). Prevention of type I diabetes in nonobese diabetic mice by virus infection. Science 239, 500–502.

Olthoff, K. M., Judge, T. A., Gelman, A. E., da Shen, X., Hancock, Turka, L. A. and Shaked, A. (1998). Adenovirus-mediated gene transfer into cold-preserved liver allografts: Survival pattern and unresponsiveness following transduction with CTLA4Ig. Nat. Med. 4, 194–200.

Parishes, N. M., Chandler, P., Quartey-Papafio, R., Simpson, E. and Cooke, A. (1993). The effect of bone marrow and thymus chimerism between nonobese diabetic (NOD) and NOD-E transgenic mice, on the expression and prevention of diabetes. Eur. J. Immunol. 23, 2667–2675.

Pennline, K. J., Roque-Gaffney, E. and Monahan, M. (1994). Recombinant human IL-10 prevents the onset of diabetes in the nonobese diabetic mouse. Clin. Immunol. Immunopathol. 71, 169–175.

Pleyer, U., Ritter, T. and Volk, H. D. (2000). Immune tolerance and gene therapy in transplantation. Immunol. Today 21, 12–14.

Pukel, C., Baquerizo, H. and Rabinovitch, A. (1988). Destruction of rat islet cell monolayers by cytokines. Synergistic interactions of interferon-gamma, tumor necrosis factor, lymphotoxin, and interleukin 1. Diabetes 37, 133–136.

Qin, L. H., Ding, Y. Z., Pahud, D. R., Robson, N. D., Shaked, A. and Bromberg, J. S. (1997). Adenovirus-mediated gene transfer of viral interleukin-10 inhibits the immune response to both alloantigen and adenoviral antigen. Hum. Gene Ther. 8, 1365–1374.

Rabinovitch, A. (1994). Immunoregulatory and cytokine imbalances in the pathogenesis of IDDM. Therapeutic intervention by immunostimulation? Diabetes 43, 613–621.

Rabinovitch, A., Suarez-Pinzon, W., Strynadka, K., Ju, Q. D., Edelstein, D., Brownlee, M., Korbutt, G. S. and Rajotte, R. V. (1999). Transfection of human pancreatic islets with an anti-apoptotic gene (bcl-2) protects beta-cells from cytokine-induced destruction. Diabetes 48, 1223–1229.

Rabinovitch, A., Suarez-Pinzon, W. L., Sorensen, O., Bleackley, R. C., Power, R. F. and Rajotte, R. V. (1995). Combined therapy with interleukin-4 and interleukin-10 inhibits autoimmune diabetes recurrence in syngeneic islet-transplanted nonobese diabetic mice. Analysis of cytokine mRNA expression in the graft.Transplantation 60, 368–374.

Rapoport, M. J., Jaramillo, A., Zipris, D., Lazarus, A. H., Serreze, D. V., Leiter, E. H., Cyopick, P., Danska, J. S. and Delovitch, T. L. (1993). Interleukin 4 reverses T cell proliferative unresponsiveness and prevents the onset of diabetes in nonobese diabetic mice. J. Exp. Med. 178, 87–99.

Robbins, P. D. and Evans, C. H. (1996). Prospects for treating autoimmune and inflammatory diseases by gene therapy. Gene Ther. *3*, 187–189.

Rosenberg, L. (1998). Clinical islet cell transplantation. Are we there yet?Int. J. Pancreatol. *24*, 145–168.

Sadelain, M. W., Qin, H. Y., Lauzon, J. and Singh, B. (1990). Prevention of type I diabetes in NOD mice by adjuvant immunotherapy. Diabetes *39*, 583–589.

Sarvetnick, N., Liggitt, D., Pitts, S. L., Hansen, S. E. and Stewart, T. A. (1988). Insulin-dependent diabetes mellitus induced in transgenic mice by ectopic expression of class II MHC and interferon-gamma. Cell *52*, 773–782.

Scharp, D. W., Lacy, P. E., Santiago, J. V., McCullough, C. S., Weide, L. G., Boyle, P. J., Falqui, L., Marchetti, P., Ricordi, C. and Gingerich, R. L. (1991). Results of our first nine intraportal islet allografts in type 1, insulin-dependent diabetic patients. Transplantation *51*, 76–85.

Schmidt-Wolf, G. D. and Schmidt-Wolf, I. G. H. (1995). Cytokines and gene therapy. Immunol. Today *16*, 173–175.

Shapiro, A. M., Lakey, J. R., Ryan, E. A., Korbutt, G. S., Toth, E., Warnock, G. L, Kneteman, N. M. and Rajotte, R. V. (2000). Islet transplantation in seven patients with type 1 diabetes mellitus using a glucocorticoid-free immunosuppressive regimen. N. Engl. J. Med. *343*, 230–238.

Shapiro, A. M., Lakey, J. R., Ryan, E. A., Korbutt, G. S., Toth, E., Warnock, G. L., Kneteman, N. M. and Rajotte, R. V. (2000). Islet transplantation in seven patients with type 1 diabetes mellitus using a glucocorticoid-free immunosuppressive regimen. N. Engl. J. Med. *343*, 230–238.

Shaw, J. A., Delday, M. I., Hart, A. W., Docherty, H. M., Maltin, C. A. and Docherty, K. (2002). Secretion of bioactive human insulin following plasmid-mediated gene transfer to non-neuroendocrine cell lines, primary cultures and rat skeletal muscle in vivo. J. Endocrinol. *172*, 653–672.

Shehadeh, N., Calcinaro, F., Bradley, B. J., Bruchlim, I., Vardi, P. and Lafferty, K. J. (1994). Effect of adjuvant therapy on development of diabetes in mouse and man. Lancet *343*, 706–707.

Short, D. K., Okada, S., Yamauchi, K. and Pessin, J. E. (1998). Adenovirus-mediated transfer of a modified human proinsulin gene reverses hyperglycemia in diabetic mice. Am. J. Physiol. *275*, E748–E756.

Smith, D. K., Korbutt, G. S., Suarez-Pinzon, W. L., Kao, D., Rajotte, R. V. and Elliott, J. F. (1997). Interleukin–4 or interleukin–10 expressed from adenovirus-transduced syngeneic islet grafts fails to prevent beta cell destruction in diabetic NOD mice. Transplantation *64*, 1040–1049.

Stevens, R. B., Lokeh, A., Ansite, J. D., Field, M. J., Gores, P. F. and Sutherland, D. E. (1994). Role of nitric oxide in the pathogenesis of early pancreatic islet dysfunction during rat and human intraportal islet transplantation. Transplant. Proc. *26*, 692.
Swenson, K. M., Ke, B. B., Wang, T., Markowitz, J. S., Maggard, M. A., Spear, G. S., Imagawa, D. K., Goss, J. A., Busuttil, R. W. and Seu, P. (1998). Fas ligand gene transfer to renal allografts in rats—Effects on allograft survival. Transplantation *65*, 155–160.
Taylor-Robinson, A. W. and Phillips, R. S. (1994). Expression of the IL-1 receptor discriminates Th2 from Th1 cloned CD4+ T cells specific for *Plasmodium chabaudi*. Immunology *81*, 216–221.
Thivolet, C., Bendelac, A., Bedossa, P., Bach, J. F. and Carnaud, C. (1991). CD8+ T cell homing to the pancreas in the nonobese diabetic mouse is CD4+ T cell-dependent. J. Immunol. *146*, 85–88.
Thule, P. M. and Liu, J. M. (2000). Regulated hepatic insulin gene therapy of STZ-diabetic rats. Gene Ther. *7*, 1744–1752.
Thule, P. M., Liu, J. and Phillips, L. S. (2000). Glucose regulated production of human insulin in rat hepatocytes. Gene Ther. *7*, 205–214.
Tiedge, M., Lortz, S., Drinkgern, J. and Lenzen, S. (1997). Relation between antioxidant enzyme gene expression and antioxidative defense status of insulin-producing cells. Diabetes *46*, 1733–1742.
Tiedge, M., Lortz, S., Munday, R. and Lenzen., S (1999). Protection against the co-operative toxicity of nitric oxide and oxygen free radicals by over-expression of antioxidant enzymes in bioengineered insulin-producing RINm5F cells. Diabetologia *42*, 849–855.
Tisch, R. and McDevitt, H. (1996). Insulin-dependent diabetes mellitus. Cell *85*, 291–297.
Tuch, B. E., Tabiin, M. T., Casamento, F. M., Simpson, A. M. and Marshall, G. M. (1998). Transplantation of genetically engineered insulin-producing hepatocytes into immunoincompetent mice. Transplant. Proc. *30*, 473.
Uchikoshi, F., Yang, Z. D., Rostami, S., Yokoi, Y., Capocci, P., Barker, C. F. and Naji, A. (1999). Prevention of autoimmune recurrence and rejection by adenovirus-mediated CTLA4Ig gene transfer to the pancreatic graft in BB rat. Diabetes *48*, 652–657.
Valera, A., Fillat, C., Costa, C., Sabater, J., Visa, J., Pujol, A. and Bosch, F. (1994). Regulated expression of human insulin in the liver of transgenic mice corrects diabetic alterations. FASEB J. *8*, 440–447.
Verdaguer, J., Yoon, J. W., Anderson, B., Averill, N., Utsugi, T., Park, B. J. and Santamaria, P. (1996). Acceleration of spontaneous diabetes in TCR-beta-transgenic nonobese diabetic mice by beta-cell cytotoxic CD8+ T cells

expressing identical endogenous TCR-alpha chains. J. Immunol. *157*, 4726–4735.
Voorbij, H. A., Jeucken, P. H., Kabel, P. J., De Haan, M. and Drexhage, H. A. (1989). Dendritic cells and scavenger macrophages in pancreatic islets of prediabetic BB rats. Diabetes *38*, 1623–1629.
Walker, R., Bone, A. J., Cooke, A. and Baird, J. D. (1988). Distinct macrophage subpopulations in pancreas of prediabetic BB/E rats. Possible role for macrophages in pathogenesis of IDDM.Diabetes *37*, 1301–1304.
Wang, B., Gonzalez, A., Hoglund, P., Katz, J. D., Benoist, C. and Mathis, D. (1998). Interleukin–4 deficiency does not exacerbate disease in NOD mice. Diabetes *47*, 1207–1211.
Warnock, G. L., Kneteman, N. M., Ryan, E., Seelis, R. E., Rabinovitch, A. and Rajotte, R. V. (1991). Normoglycaemia after transplantation of freshly isolated and cryopreserved pancreatic islets in type 1 (insulin-dependent) diabetes mellitus. Diabetologia *34*, 55–58.
Weirs, G. C., Bonner-Weir, S. and Leahy, J. L. (1990). Islet mass and function in diabetes and transplantation. Diabetes *39*, 401–405.
Wells, J. M. and Melton, D. A. (1999). Vertebrate endoderm development. Annu. Rev. Cell Dev. Biol. *15*, 393–410.
Wogensen, L., Lee, M. S. and Sarvetnick, N. (1994). Production of interleukin 10 by islet cells accelerates immune-mediated destruction of beta cells in nonobese diabetic mice. J. Exp. Med. *179*, 1379–1384.
Wood, K. J. and Fry, J. (1999). Gene thereapy: Potential applications in clinical transplantation. Expert. Rev. Mol. Med *8*, 1–20.
Yagi, H., Matsumoto, M., Kunimoto, K., Kawaguchi, J., Makino, S. and Harada, M. (1992). Analysis of the roles of CD4+ and CD8+ T cells in autoimmune diabetes of NOD mice using transfer to NOD athymic nude mice. Eur. J. Immunol. *22*, 2387–2393.
Yangs, Z. D., Rostami, S., Koeberlein, B., Barker, C. F. and Naji, A. (1999). Cardiac allograft tolerance induced by intra-arterial infusion of recombinant adenoviral CTLA4Ig. Transplantation *67*, 1517–1523.
Yasuda, H., Nagata, M., Arisawa, K., Yoshida, R., Fujihira, K., Okamoto, N., Moriyama, H., Miki, M., Saito, I., Hamada, H., Yokono, K. and Kasuga, N. (1998). Local expression of immunoregulatory IL-12p40 gene prolonged syngeneic islet graft survival in diabetic NOD mice. J. Clin. Invest. *102*, 1807–1814.
Ye, J. and Laychock, S. G. (1998). A protective role for heme oxygenase expression in pancreatic islets exposed to interleukin–1beta. Endocrinology *139*, 4155–4163.

Yoon, J. W. and Jun, H. S. (2002). Recent advances in insulin gene therapy for type 1 diabetes. Trends Mol. Med. *8*, 62–68.

Zhang, Y. C., Pileggi, A., Agarwal, A., Molano, R. D., Powers, M., Brusko, T., Wasserfall, C., Goudy, K., Zahr, E., Poggioli, R., Scott-Jorgensen, M., Campbell-Thompson, M., Crawford, J. M., Nick, H., Flotte, T., Ellis, T. M., Ricordi, C., Inverardi, L. and Atkinson, M. A. (2003). Adeno-associated virus-mediated IL–10 gene therapy inhibits diabetes recurrence in syngeneic islet cell transplantation of NOD mice. Diabetes *52*, 708–716.

Zwacka, R. M., Zhou, W. H., Zhang, Y. L., Darby, C. J., Dudus, L., Halldorson, J., Oberley, L. and Engelhardt, J. F. (1998). Redox gene therapy for ischemia/reperfusion injury of the liver reduces AP1 and NF-kappa B activation. Nat. Med. *4*, 698–704.

CHAPTER 7

Gene therapy for kidney diseases

Sifeng Chen,[1,2] Kirsten M. Madsen,[2] C. Craig Tisher[2] and Anupam Agarwal[1,2]

[1]*Department of Medicine, Nephrology Research and Training Center, University of Alabama at Birmingham, Birmingham, AL 35294, USA*
[2]*Department of Medicine, Division of Nephrology, Hypertension and Transplantation, University of Florida, Gainesville, FL 32610, USA*

Over the past decade significant advances in molecular genetics have led scientists to initiate gene therapy trials in humans and indeed, this modality has been attempted in several clinical situations, especially in the treatment of certain metabolic disorders, cancer, hypertension, and sepsis. Gene therapy has the potential to provide a therapeutic strategy for numerous renal diseases such as diabetic nephropathy, Alport syndrome, polycystic kidney disease, and inherited renal tubular disorders as well as chronic transplant rejection and acute renal failure. However, gene therapy in the kidney has not been achieved with great success as yet, due to many obstacles that have to be overcome. Most of these barriers relate to unique structure–function relationships in the kidney, requiring gene delivery into specific regions and cell types at an appropriate time that would permit transgene expression for a desirable period. The selection of vectors, methods of delivery, choice of therapeutic genes, and targeting to the specific compartment are critical issues for the success of gene therapy for kidney diseases and related entities. The purpose of this review is to furnish a discussion of the available vector systems, particularly the recombinant adeno-associated virus (rAAV) vector, methods of gene delivery to the kidney, and the potential for gene therapy as a strategy for selected renal diseases.

7.1. Structure–function correlations

The kidney is a highly vascularized organ receiving about 25% of the cardiac output. While this would seem to be a favorable factor for renal gene delivery via the vasculature, the anatomical architecture of the kidney consisting of over a million filtering units (in the human kidney) and a multitude of cell types, each with a specific function, has precluded significant advances in this area of investigation. The functional unit of the kidney is the nephron, which consists of a glomerulus and a tubular compartment enveloped by an intricate peritubular capillary network. The tubular compartment is made up of the proximal tubule, loop of Henle and the distal tubule. Although not part of the nephron, the collecting duct that continues on from the distal tubule is an important component of the renal parenchyma and a potential target for gene therapy. Each of these compartments contain specialized cells that have specific function(s) and distinct transporters, channels, and receptors on their apical or basolateral surfaces. Most kidney diseases result from dysfunction in specific regions and cell types in the nephron as a consequence of a local or systemic condition. For example, obstruction to the urinary tract (renal pelvis, ureter, bladder, or urethra) from a local condition (e.g., stone, fibrosis, cancer) leads to hydronephrosis and eventual scarring of the kidney and loss of function. Diabetes mellitus is an example of a systemic condition in which the initial kidney damage occurs in the microvasculature, specifically in the glomerulus. Unlike the heart, liver or the brain, the fact that the kidney is a paired organ is an additional barrier to gene therapy, particularly since most systemic diseases affect both kidneys. Thus, delivery of genes for most kidney diseases would require either a systemic approach or local gene delivery into both kidneys. An exception to this requirement is the transplant setting, where a single kidney is available ex vivo for genetic manipulation prior to transplantation.

7.2. Vector systems for gene delivery

Vectors carry gene(s) of interest, protect genes from being quickly destroyed by the host, facilitate entry of genes into cells, sometimes integrate genes into chromosomes (viral vectors), and facilitate gene transcription. An ideal vector should be safe in the host, specific in targeting a tissue or organ, efficient in gene transfer, controllable for the level of gene expression required, stable and reliable in gene expression and vector production. Both non-viral and viral vectors are available and will be discussed in the context of gene therapy in the kidney.

7.2.1. Non-viral vectors

7.2.1.1. DNA
Naked DNA either in the form of a cDNA or an oligonucleotide has been used for kidney gene therapy. Since the cell membrane is impermeable to DNA, its uptake is nonspecific and occurs via phagocytosis. In the kidney, tubular cells, mesangial cells, and interstitial macrophages have significant phagocytic capability. Several studies have reported the use of systemic administration of oligonucleotides in models of renal injury (Noiri et al., 1996; Haller et al., 1998; Lai et al., 1998; Chen et al., 1999; Kuemmerle et al., 2000; Tomita et al., 2000; Dai et al., 2002; Maruyama et al., 2002; Azuma et al., 2003). Most of the DNA accumulates in the renal proximal tubules (Kuemmerle et al., 2000; Noiri et al., 1996). The reason why tubular cells in other regions are not transduced is unclear. Noiri et al. (1996) used intracardiac administration of phosphorothioated antisense oligonucleotides targeting the inducible nitric oxide synthase (iNOS) gene and demonstrated selective

blockade of the iNOS gene in proximal tubules in a rat model of renal ischemia. Antisense inhibition of iNOS was associated with improvement in renal structure and function. Chen et al. (1999) perfused kidneys with unformulated antisense oligonucleotides to intercellular adhesion molecule (ICAM)-1 prior to transplantation in syngeneic nephrectomized recipients. Expression of ICAM-1 protein and mRNA was reduced and was associated with improved renal function (Chen et al., 1999). Several studies have reported efficient gene transfer, albeit transient, using naked plasmid DNA delivered by direct intraparenchymal injection, by intravenous injection via a peripheral vein or directly into the renal vein (Kuemmerle et al., 2000; Yang et al., 2001; Dai et al., 2002; Maruyama et al., 2002; Saito et al., 2003). Intravenous injection of a plasmid encoding hepatocyte growth factor (HGF) resulted in the detection of significant levels of HGF protein in mouse kidneys and ameliorated renal dysfunction in a model of folic acid-induced renal cell injury (Dai et al., 2002). Interestingly, this approach has also been used in a model of chronic renal fibrosis following unilateral ureteral obstruction in mice where significant decreases in extracellular matrix proteins were observed following delivery of HGF as a plasmid vector (Yang et al., 2001). While the results of these studies appear promising, the precise localization of the transgenes in the kidney following delivery of naked plasmid DNA requires further investigation.

Transfection efficacy of naked DNA can be increased by physical methods such as electroporation and sonication. Electroporation employs electric pulses to punch holes in the cell membrane, usually smaller than 10 nm but larger than oligonucleotides. With the use of electroporation, DNA was delivered into the cytosol of cells by diffusion. Since its introduction in 1982, in vivo transfection has been achieved in skeletal muscle, liver, skin, tumors, testis, and the kidney. Tsujie et al. (2001) developed a method to target glomeruli using electroporation in vivo wherein injection of plasmid DNA via the renal artery was followed by application of electric fields. The kidney was electroporated by sandwiching the organ between a pair of oval-shaped, tweezer-type electrodes prior to delivery of electric pulses. Subsequently, transgenes were detected in mesangial cells of the glomeruli (Tsujie et al., 2001). Sonication using ultrasound has been reported to enhance expression of transgenes in the kidney. Azuma et al. (2003) used oligonucleotides against the transcription factor, NF-κB, and demonstrated increased transduction in renal tissue by the use of ultrasound. In a rat renal allograft model, they reported improved graft function and significantly reduced expression of NF-κB-related cytokines and proinflammatory genes using this approach. Recently, Lan et al. (2003) reported the use of a doxycycline-inducible expression vector containing the Smad7 gene (an inhibitor of TGF-β signaling) to achieve transduction in the kidney using an ultrasound-microbubble system. In a rat model of unilateral ureteral obstruction, they infused plasmid DNA containing Smad7 with Optison (echocardiographic contrast microbubbles) via the renal artery while an ultrasound transducer was applied on the kidney. Transgene expression was observed in almost all glomerular cells, vascular and perivascular cells and medullary tubular and interstitial cells. Smad7 gene transfer was associated with a significant decrease in renal fibrosis, mononuclear infiltration, and tubular atrophy compared to control animals. No significant toxicity was observed. A potential drawback of this approach is the possibility of DNA fragmentation by sonication, a problem that can be avoided by complexing the DNA with cationic polymers.

7.2.1.2. Liposome-mediated gene transfer
Cationic lipids such as liposomes, micelles, emulsions, and nanoparticles, provide a lipid envelope for DNA and allow easier passage into cells through lipophilic cell membranes. The process is not cell specific but allows the transfer of large genes. When delivered through the renal artery or retrograde via the renal pelvis or ureter, gene expression occurs predominantly in renal tubular cells (Lien and Lai, 1997). No expression in glomeruli, vessels, or the

interstitial compartment has been reported. Although there is no immune response against cationic lipids, the transduction efficiency is low and gene expression is transient.

7.2.1.3. Hemagglutinating virus of Japan (HVJ; Sendai virus) with liposomes

HVJ is a member of the paramyxovirus family. HVJ-liposome, a fusigenic virosome, enables the introduction of genetic materials directly into the cytosol without degradation. In addition, the introduction of non-histone nuclear protein [high-mobility group (HMG–1)], facilitates transfer of foreign DNA to the nucleus. This system has been used successfully in the transfer of genes <100 kbp into many cell types including glomerular cells without cell injury. Tomita et al. (1992) first described the use of HVJ liposomes for renal gene delivery and observed selective transduction of glomeruli. More recently, they have used HVJ liposomes to deliver NF-κB decoy oligonucleotides in a rat model of glomerulonephritis (Tsujie et al., 2000). Significant expression was observed in glomerular cells, and treatment with the NF-κB decoy was associated with a decrease in both proteinuria and renal inflammation (Tsujie et al., 2000). Similar to other non-viral modalities of gene transfer, the HVJ-liposome method is also limited by the relatively short-term of the gene expression.

7.2.2. Viral vectors

Viral vectors usually have a higher transduction efficacy than non-viral vectors. Many viral vectors have the capability to integrate into the host genome. For example, wild type adeno-associated virus vector integrates into a selected site on chromosome 19 while rAAV and retroviral vectors integrate into the genome randomly, which is of some concern. Many AAV and retroviral vector molecules remain episomal after transduction although gene expression is prolonged. On the other hand, adenoviruses do not integrate into the host genome, but do induce adverse host immune responses.

7.2.2.1. Retrovirus

Migration of retroviral vectors from the cytoplasm into the nucleus of infected cells requires mitosis for nuclear membrane breakdown. Although retroviruses can transduce cultured renal cells in vitro easily, it is difficult to transduce renal cells in the mature kidney because the majority of kidney cells exist in a quiescent state in vivo. Only weak expression was detected in proximal tubular cells even though folic acid was used to induce mitosis of the cells (Bosch et al., 1993). Among retroviruses, lentiviruses are unique because they can infect non-dividing cells. Lentiviruses have been widely tested in neural tissue, but limited success has been achieved in the kidney. A recent study by Gusella et al. (2002) demonstrated modest transgene expression in renal tubules in mice following direct intraparenchymal or retrograde ureteral infusion. Intrarenal arterial or venous injection was associated with lower transduction efficiency (Gusella et al., 2002). Transgene expression was stable for up to 3 months and no toxicity was observed. No expression was seen in glomeruli, vasculature, or interstitial cells.

7.2.2.2. Adenovirus

Adenoviral vectors are the most efficient and popular viral vectors for transient gene expression. These vectors are also the most commonly used vehicles for gene delivery in the kidney. They are easy to prepare, can infect both dividing and non-dividing cells, and have high transduction efficiency. Depending on the route of administration and specific vector modification, adenoviral-mediated transgene expression has been reported in renal tubules, glomeruli, and the vasculature. Systemic intravenous administration of an adenoviral vector containing the aquaporin 1 gene resulted in significant transgene expression in renal proximal tubules in aquaporin 1 knockout mice (Yang et al., 2000). Interestingly, the inability to concentrate urine upon water deprivation was reversed by adenoviral-mediated aquaporin 1 gene delivery in these mice (Yang et al., 2000). Incorporation of specific ligands such as integrins and fiber-modified adenoviral vectors has led to successful transduction of the intrarenal vasculature (McDonald et al., 1999). Intense expression of adenovirus-delivered genes can be observed in the renal medulla after retrograde intrapelvic injection. Nahman et al. (2000) have used adenoviral vectors conjugated to microspheres to transduce glomeruli. In vivo perfusion of pig kidneys with a recombinant adenovirus resulted in intense and diffuse expression of the alpha5 (IV) chain in glomeruli and its deposition into the glomerular basement membrane (Heikkila et al., 2001). However, a significant drawback with the use of adenoviral vectors is the propensity for adverse host immune responses. Several modifications have been made in the adenoviral proteins to eliminate or reduce immune responses.

7.2.2.3. Adeno-associated virus

The adeno-associated viruses are members of the parvovirus family. There are eight known serotypes of rAAV that infect primate cells, designated AAV1-8. Thus far AAV serotype 2 (AAV2) has been the most extensively studied serotype. rAAV has been used in several animal and human clinical trials in diseases such as cystic fibrosis (Flotte et al., 1993), alpha-1-antitrypsin deficiency (Song et al., 1998), and hemophilia B (Kay et al., 2000). rAAV vectors have several distinct advantages over other gene delivery vectors because rAAV results in long-term transgene expression and infects both dividing and non-dividing cells with no significant side effects, particularly with respect to immune responses. A drawback of AAV vectors is the size limitation for packaging (<5 kb), which restricts the use of these vectors to the insertion of small genes. However, recent advances in vector engineering of AAV may be able to overcome this barrier. Infection with wild-type AAV alone leads to the establishment of a long-term latency which is primarily due to site-specific integration in the AAVS1 site on human chromosome 19, although some forms can persist as episomal forms or are integrated at other sites (Giraud et al., 1994; Berns and Linden, 1995).

rAAV-mediated gene delivery resulting in long-term expression has been reported in a wide variety of tissues including retina (Flannery et al., 1997), lung (Flotte et al., 1993), muscle (Song et al., 1998; Rudich et al., 2000; Takahashi et al., 2002), liver (Snyder et al., 1997), brain (McCown et al., 1996), spinal cord (Peel et al., 1997), and pancreatic islets (Flotte et al., 2001). However, limited experience has been reported with the use of rAAV in the kidney. Lipkowitz et al. (1999) injected rAAV carrying green fluorescent protein (GFP) and/or β-galactosidase directly into the renal parenchyma of mouse kidneys and demonstrated transduction of renal tubular cells only along the needle track. While these results were encouraging, the disadvantage of the direct intra-parenchymal approach as well as the limited transduction observed, prompted us to evaluate other delivery strategies in the kidney. Following intrarenal arterial delivery of a rAAV2-GFP vector we observed successful transduction of renal tubular epithelial cells (Chen et al., 2003). Proximal tubule cells, specifically in the S_3 segment, as well as intercalated cells in the medullary collecting duct were transduced. No transduction was observed in blood vessels, glomeruli, or the interstitium. The ischemia associated with the procedure of gene delivery caused mild changes of tubular injury and interstitial mononuclear cell infiltration and was observed in both saline and vector-treated animals, suggesting that these morphological changes were not due to an adverse immune response to the viral vector or the transgene.

The reason(s) observed in our studies for the preferential transduction of the proximal tubule and intercalated cells is not entirely clear. Several possible mechanisms may be implicated. The early steps of AAV infection involve attachment to a variety of cell surface receptors (heparan sulfate proteoglycan, fibroblast growth factor receptor, and α_v-β_5 integrin) followed by a clathrin-dependent or independent internalization process (Summerford and Samulski, 1998; Qing et al., 1999; Summerford et al., 1999). AAV2 generally requires a helper virus (usually an adenovirus or herpes virus) or a physical or chemical insult to undergo second strand synthesis and productive replication in vitro and in vivo (Ferrari et al., 1996). It is possible that ischemic cell injury results in increased surface expression of receptors that facilitate AAV uptake in the proximal tubule, particularly in the S_3 segment, a region that is more susceptible to ischemic injury (Lieberthal and Nigam, 1998). Selective transduction of intercalated cells in the collecting duct was also observed. Because these cells are capable of apical endocytosis (Brown et al., 1987) it is tempting to speculate that the vector might have gained access to the tubule lumen and subsequently been absorbed by endocytosis. We are currently evaluating expression of transgenes using the different AAV serotypes and capsid mutants as well as specific promoter systems to optimize cell-specific gene delivery in the kidney.

7.2.2.4. Chimeric viral vectors
The development of chimeric viral vectors offers unique features that would combine the advantages of two or more different viral vectors while avoiding their disadvantages. For example, the chimera of adenovirus and AAV vectors would transduce a variety of dividing and non-dividing cells, carry a large insert gene, integrate into a specific host genome site, be easily produced and lack the ability to induce adverse immune responses (Shayakhmetov et al., 2002). The successful use of these vectors for renal gene delivery needs further investigation.

7.3. Methods of gene delivery

Besides a systemic approach for the delivery of a therapeutic gene product to the kidney, the anatomical structure of this organ permits several different approaches for gene delivery. These include intrarenal artery, intrarenal vein, direct intraparenchymal, retrograde intrapelvic, and subcapsular injection. The selection of any of these routes of administration is dependent on several factors such as (i) location of the cells to be targeted, (ii) whether the gene product is secreted, (iii) whether the disease is specific to both kidneys, a single kidney or multiple organs and, (iv) the consequences of gene delivery to the wrong cell type in the kidney.

7.3.1. Gene delivery into tissue outside the kidney

This approach can provide secretory and soluble bioactive factors that will have effects on the kidney through the blood circulation. The gene product will affect not only the kidneys but other organs as well. It is a suitable approach when the renal disorder occurs as part of a multiple organ disease such as diabetes mellitus or systemic lupus erythematosus. It is also an appropriate modality to supplement the endocrine functions of the kidney. For example, the normal kidney synthesizes and secretes erythropoietin into the blood stream. By delivering the erythropoietin gene into tissues other than the kidney, anemia secondary to renal disease can be treated. The most commonly used tissue for such purposes is the skeletal muscle since it is easily accessible and an efficient platform for producing secreted proteins. Recent studies have utilized AAV vectors encoding erythropoietin targeted to the skeletal muscle in monkeys and have reported significant levels of erythropoietin in the circulation.

7.3.2. Intrarenal artery or intrarenal vein infusion

The gene of interest is delivered throughout the kidney via the blood circulation. Theoretically, this appears to be an ideal approach if gene-transfer to the entire kidney is desired. However, due to the specific mechanisms of uptake of different vectors and the architecture of the kidney, this has not been accomplished. This approach is well suited to the transplant setting where the donor kidney can be transduced ex vivo prior to implantation into a recipient.

7.3.3. Intraparenchymal or subcapsular injection

Vectors can be injected into the kidney at the location of choice or at multiple sites either beneath the capsule or into the renal parenchyma. The disadvantage of this technique is that the transgene expression is limited to renal tissue surrounding the needle track and the needle penetration per se can cause cell damage. This approach could be relevant for localized renal tumors where anti-angiogenic, pro-apoptotic, or tumor-suppressive genes could be injected.

7.3.4. Retrograde ureteral or pelvic injection

Gene cassettes can be delivered into the kidney through a retrograde approach via the lower urinary tract. This is a good approach if tubular cells are the only cell types that need to be transduced and is clinically applicable given the relatively less invasive access to the lower urinary system by cystoscopy.

7.4. Targeting specific cells in the kidney

7.4.1. Vascular cells and glomeruli

Vascular cells play a pivotal role in inflammatory renal disease, renal allograft rejection, hypertension, and diabetic nephropathy. However, endothelial and smooth muscle cells are relatively resistant to gene transfer. The only vector that has been reported to successfully transduce the intrarenal vasculature is the adenovirus. McDonald et al. (1999) utilized a fiber-modified adenoviral vector containing the RGD integrin-binding motif and observed selective transgene expression in the renal cortical vasculature. Ye et al. (2001, 2002) demonstrated that rat glomerular endothelial cells could be efficiently transduced by slowly infusing a recombinant adenovirus (Ad.CBlacZ) into the right renal artery for a period of 15 min. High levels of lacZ expression were achieved in renal glomeruli, including occasional expression in endothelial cells. These authors have developed two techniques involving "portal vein clamping" and prolonged renal infusion of adenoviral vectors and observed transgene expression in glomeruli of mice and rats (Ye et al., 2002). In the "portal vein clamping" method that was used in mice, the portal triad including the hepatic artery, portal vein, and bile duct were temporarily clamped for 30 min just below their entry into the liver. During this time, adenoviral vectors were injected into the venous system. Bypassing the liver resulted in high levels of transgene expression in glomerular capillaries, mostly in glomerular endothelial cells (Ye et al., 2002). In the prolonged renal infusion method, the right kidney was selectively perfused with adenoviral vectors through the superior mesenteric artery while the kidney was cooled with ice to limit ischemic injury (Ye et al., 2002). Transgene expression was observed in glomeruli, specifically in glomerular endothelial cells. No expression was seen in the opposite kidney or in other organs.

Tomita et al. (2002) demonstrated that systemic delivery of HVJ-liposomes coupled with OX-7 results in efficient transfer of oligoucleotides in rat glomeruli. Genes can also be delivered into glomeruli by transfusion of modified cells. Injection of mesangial cells transfected with the beta-galactosidase gene into the renal artery resulted in the trapping of transduced cells in the capillary lumen of glomeruli and transgene expression for 4 weeks (Tomita et al., 2002). Intraarterial administration of microsphere conjugated adenoviral vectors has been demonstrated to selectively transduce glomeruli (Nahman et al., 2000).

In our attempts to achieve gene delivery in vascular endothelial cells in vitro and in vivo, we have observed significantly higher transduction using rAAV serotype 1 and 5 vectors (Chen et al., 2005). Other investigators have incorporated endothelial-specific ligands such as SIGYPLP in AAV capsids, and reported increased transduction efficiency in endothelial cells in culture (Nicklin et al., 2001).

7.4.2. Tubular epithelial cells

Tubular epithelial cells (TEC) lining the nephron consist of more than 10 distinct subtypes each with a specific function. In most disease states, the gene defect involves a particular TEC subtype. Thus, theoretically, specific TEC-targeted gene delivery could offer an exciting therapeutic avenue. The use of ligands or cell-specific promoters would enable selective transduction of affected TEC. TEC have been transduced in vitro, ex vivo, and in vivo. Studies have reported successful transgene expression in renal tubular cells using non-viral and viral vectors. The approaches used for in vivo delivery of genes into TEC include the renal artery, peripheral veins, the renal pelvis, and direct intraparenchymal injection. In our studies using adeno-associated viral vectors, we were able to achieve transgene expression in the S_3 segment of the proximal tubule and in intercalated cells (Chen et al., 2003). Although the mechanism of the selective transduction is unknown, specific receptor-ligand binding between vector and cell may be involved. Lipkowitz et al. (1999) also successfully delivered AAV into TEC using direct intraparenchymal injection, although expression was limited to areas adjacent to the needle track. Retrovirus is able to infect TEC, however, transduction occurs only in the presence of a cytotoxic agent. Rappaport et al. (1995) detected oligonucleotides in the proximal TEC 30 min after intravenous injection. By injection into the renal artery as well as the intrarenal pelvis, Lien and Lai (1997) were able to transfect TEC with lipofectin-DNA complexes and observed differences in the location of the transgene depending on the route of administration of the vector DNA. The reporter gene was expressed in both the outer medulla and cortex after intrarenal artery injection, while gene expression was limited to the outer medulla after intrapelvic administration.

7.4.3. Interstitial cells

Tubulointerstitial inflammation occurs virtually in all renal diseases and is a major determinant in the progression of chronic kidney disease. There are many distinct interstitial cell types, several of which contribute to normal renal physiology and function. For example, specific types of interstitial cells in the kidney are responsible for the production and secretion of erythropoietin. Interstitial cells provide the intrarenal tubular and vascular architecture with necessary support by the production of extracellular matrix components. In addition, during inflammatory conditions a variety of immune cells from the circulation also infiltrate the kidney to expand the existing population of resident interstitial cells. Targeting genes with antifibrotic or immune modulatory function to renal interstitial cells would have significant clinical relevance in patients with renal disease. However, limited progress has been made due to the difficulty in accessing these cell types. Tsujie et al. (2000) developed an interstitial fibroblast targeting gene transfer technique using an artificial viral envelope-type HVJ-liposome method. LacZ gene or TGF-β1 expression was observed mainly in fibroblasts in the tubulointerstitial compartment. The use of vasodilators during adenoviral gene delivery has also been reported to facilitate targeting of interstitial cells (Zhu et al., 1996). Systemic gene therapy to produce secreted proteins that affect the renal interstitium is also a feasible approach to target this otherwise inaccessible region of the kidney.

7.5. Application of gene therapy for specific kidney diseases

7.5.1. Genetic renal disorders

With the understanding of the genetic basis for several inherited renal disorders, corrective gene therapy for such diseases is an exciting possibility. The molecular basis for renal diseases such as Bartter syndrome, congenital nephrotic syndrome, Fabry disease, Alport syndrome, and polycystic kidney disease as well as others has been elucidated. Inactivating mutations of specific ion transporters in the nephron results in Bartter syndrome (Kamel et al., 2002). Mutations of the nephrin gene are associated with the Finnish type of the congenital nephrotic syndrome. Mutations of podocin or alpha-actinin-4 results in congenital focal segmental glomerulosclerosis (Pollak, 2003). Fabry disease is an X-linked lysosomal storage disorder that is the result of a deficiency of the enzyme alpha-galactosidase A, which leads to the deposition of glycosphingolipids in multiple organs including the kidney (Breunig et al., 2003). Several gene mutations have been linked to renal cysts in humans, including PKD1 and PKD2 (autosomal dominant polycystic kidney disease, ADPKD), PKHD1 (autosomal recessive polycystic kidney disease, ARPKD), TSC1 and TSC2 (tuberous sclerosis), NPHP1 (nephronophthisis type I), and VHL (von-Hippel-Lindau syndrome) (Watnick and Germino, 1999). Renal replacement therapy in the form of dialysis or kidney transplantation is often required for end-stage renal failure caused by most of these genetic renal diseases. Alternative therapeutic strategies such as the use of gene therapy, particularly early in the course of these genetic diseases, would offer significant benefits. Progress has been quite encouraging in this area of experimentation.

Mutations of type IV collagen in the glomerular basement membrane result in Alport syndrome, which causes progressive loss of renal function requiring dialysis or transplantation (Kashtan, 1999). Gene therapy strategies aimed at delivery of the normal type IV collagen gene to the glomerulus for the treatment of Alport syndrome are being investigated. Heikkila et al. (2001) reported that in vivo perfusion of pig kidneys with a recombinant adenovirus encoding the alpha 5 chain of type IV collagen resulted in expression of the transgene in glomerular basement membranes as shown by in situ hybridization and immunohistochemistry. Current studies are evaluating this approach in the Samoyed dog model of Alport syndrome.

Systemic and vascular accumulation of glycosphingolipids due to mutations of the lysosomal α-galactosidase A (AGA) enzyme causes Fabry disease (Breunig et al., 2003). In the kidney, such accumulation occurs in vascular endothelial cells in the intrarenal vasculature, in tubular epithelial cells, and in all glomerular cells including podocytes, and mesangial and endothelial cells. Studies with enzyme replacement therapy for this disorder (recently approved for clinical use by the FDA in the US) have shown dramatic results. However, a drawback of this approach is the requirement for repeated administration of the enzyme, a feature that can be overcome by the use of gene therapy. In addition, for reasons not completely understood, podocyte accumulation of glycosphingolipids is not entirely cleared by enzyme replacement therapy (Nakao et al., 2003). Park et al. (2003) have recently described the use of a single intravenous injection of a rAAV encoding the human AGA gene in a mouse model of Fabry disease. Stable expression of AGA was achieved in the liver, kidney, heart, spleen, small intestine, lung, and brain of these mice and persisted for >6 months. The levels of glycosphingolipids were significantly reduced in the liver, heart, and spleen and to a lesser extent in the kidney, lung, and small intestine. Takahashi et al. (2002) demonstrated that injection of a rAAV vector containing the AGA gene into the right quadriceps muscle of Fabry knockout mice increased AGA activity in plasma to approximately 25% of normal mice. The AGA activity in various organs of treated Fabry mice remained 5–20% of those observed in normal mice and persisted for up to 30 weeks without development of AGA antibodies. Accumulated glycosphingolipids in organs of Fabry knockout mice were completely cleared within 25 weeks after vector injection (Takahashi et al., 2002). Studies with delivery of these vectors into the kidney, particularly targeted to the glomerular podocytes would be of significant interest.

Successful correction of inherited renal diseases including carbonic anhydrase II (Lai et al., 1998) and aquaporin-1 deficiency (Yang et al., 2000) have been reported. The deficiency of carbonic anhydrase II leads to renal tubular acidosis, osteopetrosis, growth retardation, and intracranial calcification. The enzyme is normally localized in the proximal tubule, the loop of Henle, and in intercalated cells in the collecting duct and is responsible for urinary acidification. Mice deficient in carbonic anhydrase II are unable to acidify their urine after an acid load. Lai et al. (1998) utilized cationic liposomes to deliver carbonic anhydrase II to mice deficient in this enzyme by retrograde intrarenal pelvic injection and demonstrated transgene expression in tubular cells of the outer medulla and at the corticomedullary junction. Transient improvement in urinary acidification after ammonium chloride treatment was observed in these studies. Yang et al. (2000) reported partial correction of the urinary concentrating defect in response to water deprivation in aquaporin-1 deficient mice by treatment with an adenoviral vector containing aquaporin-1. Aquaporin-1 is a water channel protein and is localized in the proximal tubule, descending thin limb of Henle and descending vasa recta. Aquaporin-1 deficient mice are unable to concentrate their urine during water deprivation. In the studies by Yang et al. (2000), the vector was delivered by a single tail vein injection and resulted in transgene expression in proximal tubules and medullary vasa recta. However, aquaporin-1 expression and the functional effects were lost over 3–5 weeks.

Our studies using rAAV vectors for gene delivery in the kidney (Chen et al., 2003) are likely to be applicable to the treatment of renal tubular disorders where progress with respect to gene therapy strategies has been hampered by the transient nature of gene expression. The demonstration of transgene expression in intercalated cells using rAAV (Chen et al., 2003) and the ability of rAAV to provide long-term persistence of transgenes following a single application (Flotte et al., 1993; Xiao et al., 1996; Snyder et al., 1997; Song et al., 1998; Kay et al., 2000), make these vectors ideal candidates for investigation in renal tubular disorders. While our studies have evaluated transgene expression up to 6 weeks, we anticipate that long-term persistent transgene expression will be possible with rAAV vectors, as has been reported previously in other tissues (Flotte et al., 1993; Xiao et al., 1996; Snyder et al., 1997; Song et al., 1998; Kay et al., 2000). It is possible that further modifications in vector design using specific promoters, alternate serotypes, or capsid modifications would enhance transduction efficiency.

7.5.2. Glomerular diseases

Gene therapy has the potential to interrupt the pathways leading to glomerular injury from metabolic derangements such as diabetes mellitus, and inflammatory or immune-mediated renal diseases such as systemic lupus erythematosus, as well as others. Both systemic and local approaches are applicable. Most studies have used molecules that have effects on glomerular structures including mesangial cells, glomerular basement membrane, or the capillaries. Growth factors, protein kinases, specific receptors for advanced glycosylation end products, anti-oxidants and anti-fibrotic genes represent a few of the important molecules with therapeutic potential in glomerular disease secondary to diabetes mellitus. Maintenance of normoglycemia using a gene therapy approach to produce insulin has been attempted in animal models and would be ideal to prevent the major end-organ complications of diabetes mellitus (Xu et al., 2003).

Among the various primary and secondary glomerular diseases, diabetic nephropathy represents the most common cause of stage 5 chronic kidney disease accounting for about 44% of all patients on dialysis. One important mediator of glomerular injury in diabetic nephropathy is TGF-β, which promotes the deposition of extracellular matrix leading to renal fibrosis (Bottinger and Bitzer, 2002). Transfer of a cDNA-encoding decorin, an inhibitor of TGF-β, into the skeletal muscle has been shown to ameliorate glomerular injury in a rat model of anti-Thy-1 antibody-induced experimental glomerulonephritis (Isaka et al., 1996). In separate experiments, a chimeric TGF-βRII/Fc gene, which was generated by the fusion of the ligand-binding domain of the TGF-β type II receptor and the Fc region of IgG heavy chain, was transfected to the gluteal muscle of nephritic rats using HVJ-liposome-mediated gene transfer (Isaka et al., 1999). The gene product neutralized the effects of TGF-β and reduced extracellular matrix accumulation. The inhibitory Smad protein, Smad7, would be another candidate gene that could potentially reverse the pro-fibrotic effects of TGF-β in diabetic nephropathy (Bottinger and Bitzer, 2002). As discussed earlier, Smad7 has been shown to block fibrosis in the kidney following obstructive uropathy (Terada et al., 2002; Lan et al., 2003) and in a model of bleomycin-induced lung injury (Nakao et al., 1999).

Recent studies have investigated the effects of adenoviral-mediated delivery of the adrenomedullin gene (a potent vasodilator) in streptozotocin-induced diabetic rats (Dobrzynski et al., 2002). A single tail-vein injection of the vector resulted in detectable levels of adrenomedullin in the plasma and urine of diabetic rats and was associated with a significant reduction in glycogen accumulation and tubular damage (Dobrzynski et al., 2002).

Studies by Kitamura et al. (1994) have devised a gene delivery system using modified mesangial cells. Rat mesangial cells were stably transfected with a reporter gene and transferred into the rat kidney via the renal artery. Selective entrapment occurred within the glomeruli and sustained transgene expression was observed up to 4 weeks. Glomerular injury resulted in a significantly higher transduction efficiency (Kitamura et al., 1994), suggesting that site-specific delivery of genes was feasible in glomeruli using this approach. Recent studies have used genetically-modified bone marrow cells obtained from male mice transduced with an interleukin 1 (IL-1) receptor antagonist (IL-1Ra) using a retrovirus vector (Yokoo et al., 2001). The transduced cells were infused into sublethally irradiated female recipients. Eight weeks following bone marrow transplantation, transgene expression was confirmed by the presence of the male Y antigen in bone marrow, liver, and spleen. Further, the authors demonstrated recruitment of transplanted donor cells into the glomerulus following induction of anti-glomerular basement membrane nephrotoxic serum nephritis. Renal function and histology were better preserved in the mice reconstituted with IL-1Ra-producing cells compared with mock-transduced cells. These results suggest that in addition to mesangial cells, modified bone marrow cells may also provide a vehicle for glomerular gene delivery.

Higuchi et al. (2003) have recently described a hydrodynamic-based technique to deliver viral interleukin (vIL)-10 in a rat model of crescentic glomerulonephritis. Three hours following induction of glomerulonephritis with a rabbit polyclonal anti-rat glomerular basement membrane antibody, a large volume (15 ml or 1/10th of the body weight) of plasmid DNA expressing vIL-10 solution was rapidly (~10 s) injected into the tail vein. Transgene expression was localized mainly to the liver and resulted in high circulating levels of vIL-10 which was associated with a decrease in crescent formation, inflammatory infiltration, and proteinuria.

7.5.3. Acute renal failure

Acute renal failure secondary to ischemia-reperfusion or nephrotoxins represents a major cause of morbidity and mortality in hospitalized patients, particularly in the intensive care unit setting. The proximal tubule region of the nephron suffers the most damage in acute renal injury and is therefore the target site of therapeutic interventions. While several experimental therapies have been attempted to prevent or hasten recovery from acute renal injury, the treatment is still mainly one of supportive care and dialysis when indicated. There are many situations where preventive strategies to minimize injury to renal tubules could offer a novel approach, particularly when the likelihood of acute renal failure is high. For example, patients with diabetes mellitus have a high risk of acute renal failure following administration of radiocontrast agents. At least a third of cancer patients receiving cisplatin chemotherapy develop dose-related nephrotoxicity. It is possible to increase the inherent resistance of the proximal tubules to injury by upregulating cytoprotective genes prior to exposure to nephrotoxins in high-risk settings. Examples of candidate genes include antiapoptotic and antioxidant genes such as heme oxygenase-1, superoxide dismutase, Bcl2 or anti-inflammatory cytokine genes such as interleukin-10, or growth factors such as epidermal growth factor and HGF.

Recent results with the use of HGF, delivered as a naked plasmid DNA, in a model of folic acid-induced acute renal failure are promising (Dai et al., 2002). In these studies, a single intravenous injection of a plasmid containing HGF led to the production of significant levels of human HGF protein in mouse kidneys which was associated with accelerated recovery from folate-induced acute renal injury and protected renal epithelial cells from both apoptosis and necrosis suggesting that HGF gene therapy may represent a novel therapy in acute renal failure. Yin et al. (2001) delivered an adenoviral vector containing copper-zinc superoxide dismutase in a rat model of ischemia-reperfusion. They detected about a 2.5-fold increase in superoxide dismutase activity in the kidneys of animals receiving the vector and observed improvement in renal structure and function, suggesting that superoxide dismutase was protective in this model of renal injury.

Most of these studies have used constitutively active promoter systems in the vectors for gene activation. This approach could have a potential disadvantage because continuous production of a protein by the transgene may have adverse effects. In this regard, the development of novel regulatable vector systems that would respond to physiological or pathophysiological signals to switch them on or off would be of great interest. One such "vigilant vector" has recently been described by Phillips et al. (2002). These investigators developed an rAAV vector that contains the oxygen-sensitive domain from hypoxia-inducible factor (HIF-1) and is inducible by decreases in oxygen tension. The vector system was shown to be functional in myocardial ischemia, and this vector system can be adapted to models of renal ischemia as well. Because changes in the endogenous tissue of interest would cause gene activation, unnecessary side effects from constant overexpression of the transgene can be avoided. In addition, this system will not require exogenous administration of agents such as tetracycline or rapamycin to regulate expression levels of the transgene.

7.5.4. Transplantation

Ischemia-reperfusion injury leading to delayed graft function and immune-mediated processes that cause acute and chronic rejection are major barriers to the proper functioning of renal allografts. The availability of the kidney for ex vivo manipulation prior to transplantation is a great advantage for gene therapy because it obviates the potential adverse consequences of systemic exposure to the vector as well as the requirement for systemic immunosuppression. Genetic modulation of kidney grafts has focused on reducing graft immunogenecity, induction of tolerance, and overexpression of cytoprotective genes such as interleukin-10, CTLA4Ig, heme oxygenase-1 and others (Chen et al., 1999; Tomasoni et al., 2000; Azuma et al., 2003; Blydt-Hansen et al., 2003). Azuma et al. (2003) used ultrasound-Optison-mediated gene delivery in donor kidneys and demonstrated significant prolongation of graft survival by NF-κB-decoy oligonucleotides in a rat renal allograft model. Blydt-Hansen et al. (2003) have recently evaluated the effects of adenoviral-heme oxygenase-1 gene delivery in rat renal isografts. Transgene expression was limited to renal tubules and was associated with better preserved renal function and histology in the animals receiving Ad-heme oxygenase-1 compared to those receiving Ad-beta-galactosidase or phosphate buffered saline.

Studies by Tomasoni et al. (2000) have demonstrated prolonged survival of rat renal allografts using adenovirus-mediated delivery of the CTLA4Ig, the gene product that blocks T cell activation. Hammer et al. (2002) have described a novel approach to obtain antigen-dependent gene expression in kidney allografts. Using the Moloney murine leukemia virus-based enhanced green fluorescence protein in a retroviral vector, they modified rat T lymphocytes ex vivo and transferred them into rats that received syngeneic, allogenic, or third party kidney transplants. It was demonstrated that GFP-positive T lymphocytes were activated and accumulated only in allografts but not in syngeneic or third party transplants. This approach of antigen-specific gene targeting using ex vivo-modified cells could provide an innovative strategy for successful gene therapy in organ transplantation.

7.6. Summary

The strategies for gene delivery for application in renal diseases depend on the nature of the underlying disease as well as on the region of the kidney and cell types involved. The unique structure–function relationships and the multiple cell types present in the kidney are major barriers to the ultimate success of gene delivery to treat kidney diseases. The development of vectors to target specific cell types in the kidney and the design of regulatable expression systems would be of great interest. Among the available gene delivery systems, rAAV vectors have several distinct advantages. rAAV has a broad host range and is capable of transducing both dividing and non-dividing cells. rAAV results in long-term transgene expression following a single application and infects cells with no significant side effects, particularly with respect to immune responses. While gene therapy represents a promising strategy and has significant therapeutic potential in kidney diseases, the presently available technology and vector systems are not completely efficacious and require further investigation prior to clinical use in patients.

Acknowledgments

This work was supported by grants from the Juvenile Diabetes Research Foundation (JDRF) Gene Therapy Center grant for the Prevention of Diabetes and Diabetic Complications at the University of Florida and the University of Miami, and NIH grant R21 DK067472 (to A.A.).

References

Azuma, H., Tomita, N., Kaneda, Y., Koike, H., Ogihara, T., Katsuoka, Y. and Morishita, R. (2003). Transfection of NFkappaB-decoy oligodeoxynucleotides using efficient ultrasound-mediated gene transfer into donor kidneys prolonged survival of rat renal allografts. Gene Ther. *10*, 415–425.

Berns, K. I. and Linden, R. M. (1995). The cryptic life style of adeno-associated virus. Bioessays *17*, 237–245.

Blydt-Hansen, T. D., Katori, M., Lassman, C., Ke, B., Coito, A. J., Iyer, S., Buelow, R., Ettenger, R., Busuttil, R. W. and Kupiec-Weglinski, J. W. (2003). Gene transfer-induced local heme oxygenase-1 overexpression protects rat kidney transplants from ischemia/reperfusion injury. J. Am. Soc. Nephrol. *14*, 745–754.

Bosch, R. J., Woolf, A. S. and Fine, L. G. (1993). Gene transfer into the mammalian kidney: Direct retrovirus-transduction of regenerating tubular epithelial cells. Exp. Nephrol. *1*, 49–54.

Bottinger, E. P. and Bitzer, M. (2002). TGF-beta signaling in renal disease. J. Am. Soc. Nephrol. *13*, 2600–2610.

Breunig, F., Weidemann, F., Beer, M., Eggert, A., Krane, V., Spindler, M., Sandstede, J., Strotmann, J. and Wanner, C. (2003). Fabry disease: Diagnosis and treatment. Kidney Int. *84(Suppl.)*, S181–S185.

Brown, D., Weyer, P. and Orci, L. (1987). Nonclathrin-coated vesicles are involved in endocytosis in kidney collecting duct intercalated cells. Anat. Rec. *218*, 237–242.

Chen, S., Agarwal, A., Glushakova, O. Y., Jorgensen, M. S., Salgar, S. K., Poirier, A., Flotte, T. R., Croker, B. P., Madsen, K. M., Atkinson, M. A., Hauswirth, W. W., Berns, K. I. and Tisher, C. C. (2003). Gene delivery in renal tubular epithelial cells using recombinant adeno-associated viral vectors. J. Am. Soc. Nephrol. *14*, 947–958.

Chen, S., Kapturczak, M. H., Loiler, S. A., Zolotukhin, S., Glushakova, O. Y., Madsen, K. M., Samulski, R. J., Hauswirth, W. W., Berns, K. I., Flotte, T. R., Atkinson, M. A., Tisher, C. C. and Agarwal, A. (2005). Efficient transduction of vascular endothelial cells with recombinant adeno-associated virus serotype 1 and 5 vectors. Hum. Gene Ther. *16*, 235–247.

Chen, W., Bennett, C. F., Wang, M. E., Dragun, D., Tian, L., Stecker, K., Clark, J. H., Kahan, B. D. and Stepkowski, S. M. (1999). Perfusion of kidneys with unformulated "naked" intercellular adhesion molecule-1 antisense oligodeoxynucleotides prevents ischemic/reperfusion injury. Transplantation *68*, 880–887.

Dai, C., Yang, J. and Liu, Y. (2002). Single injection of naked plasmid encoding hepatocyte growth factor prevents cell death and ameliorates acute renal failure in mice. J. Am. Soc. Nephrol. *13*, 411–422.

Dobrzynski, E., Montanari, D., Agata, J., Zhu, J., Chao, J. and Chao, L. (2002). Adrenomedullin improves cardiac function and prevents renal damage in streptozotocin-induced diabetic rats. Am. J. Physiol. Endocrinol. Metab. *283*, E1291–E1298.

Ferrari, F. K., Samulski, T., Shenk, T. and Samulski, R. J. (1996). Second-strand synthesis is a rate-limiting step for efficient transduction by recombinant adeno-associated virus vectors. J. Virol. *70*, 3227–3234.

Flannery, J. G., Zolotukhin, S., Vaquero, M. I., La Vail, M. M., Muzyczka, N. and Hauswirth, W. W. (1997). Efficient photoreceptor-targeted gene expression in vivo by recombinant adeno-associated virus. Proc. Natl. Acad. Sci. USA *94*, 6916–6921.

Flotte, T. R., Afione, S. A., Conrad, C., McGrath, S. A., Solow, R., Oka, H., Zeitlin, P. L., Guggino, W. B. and Carter, B. J. (1993). Stable in vivo expression of the cystic fibrosis transmembrane conductance regulator with an adeno-associated virus vector. Proc. Natl. Acad. Sci. USA *90*, 10613–10617.

Flotte, T., Agarwal, A., Wang, J., Song, S., Fenjves, E., Inverardi, L., Chesnut, K., Afione, S., Loiler, S., Wasserfall, C., Kapturczak, M., Ellis, T., Nick, H. S. and Atkinson, M. (2001). Efficient ex vivo transduction of pancreatic islet cells with recombinant adeno-associated virus vectors. Diabetes *50*, 515–520.

Giraud, C., Winocour, E. and Berns, K. I. (1994). Site-specific integration by adeno-associated virus is directed by a cellular DNA sequence. Proc. Natl. Acad. Sci. USA *91*, 10039–10043.

Gusella, G. L., Fedorova, E., Hanss, B., Marras, D., Klotman, M. E. and Klotman, P. E. (2002). Lentiviral gene transduction of kidney. Hum. Gene Ther. *13*, 407–414.

Haller, H., Maasch, C., Dragun, D., Wellner, M., von Janta-Lipinski, M. and Luft, F. C. (1998). Antisense oligodesoxynucleotide strategies in renal and cardiovascular disease. Kidney Int. *53*, 1550–1558.

Hammer, M. H., Schroder, G., Risch, K., Flugel, A., Volk, H. D., Lehmann, M. and Ritter, T. (2002). Antigen-dependent transgene expression in kidney transplantation: A novel approach using gene-engineered T lymphocytes. J. Am. Soc. Nephrol. *13*, 511–518.

Heikkila, P., Tibell, A., Morita, T., Chen, Y., Wu, G., Sado, Y., Ninomiya, Y., Pettersson, E. and Tryggvason, K. (2001). Adenovirus-mediated transfer of type IV collagen alpha5 chain cDNA into swine kidney in vivo: Deposition of the protein into the glomerular basement membrane. Gene Ther. *8*, 882–890.

Higuchi, N., Maruyama, H., Kuroda, T., Kameda, S., Iino, N., Kawachi, H., Nishikawa, Y., Hanawa, H., Tahara, H., Miyazaki, J. and Gejyo, F. (2003). Hydrodynamics-based delivery of the viral interleukin-10 gene suppresses experimental crescentic glomerulonephritis in Wistar-Kyoto rats. Gene Ther. *10*, 1297–1310.

Isaka, Y., Akagi, Y., Ando, Y., Tsujie, M., Sudo, T., Ohno, N., Border, W. A., Noble, N. A., Kaneda, Y., Hori, M. and Imai, E. (1999). Gene therapy by transforming growth factor-beta receptor-IgG Fc chimera suppressed extracellular matrix accumulation in experimental glomerulonephritis. Kidney Int. *55*, 465–475.

Isaka, Y., Brees, D. K., Ikegaya, K., Kaneda, Y., Imai, E., Noble, N. A. and Border, W. A. (1996). Gene therapy by skeletal muscle expression of decorin prevents fibrotic disease in rat kidney. Nat. Med. *2*, 418–423.

Kamel, K. S., Oh, M. S. and Halperin, M. L. (2002). Bartter's, Gitelman's, and Gordon's syndromes. From physiology to molecular biology and back, yet still some unanswered questions. Nephron. *92(Suppl. 1)*, 18–27.

Kashtan, C. E. (1999). Alport syndrome. An inherited disorder of renal, ocular, and cochlear basement membranes. Medicine (Baltimore) *78*, 338–360.

Kay, M. A., Manno, C. S., Ragni, M. V., Larson, P. J., Couto, L. B., McClelland, A., Glader, B., Chew, A. J., Tai, S. J., Herzog, R. W., Arruda, V., Johnson, F., Scallan, C., Skarsgard, E., Flake, A. W. and High, K. A. (2000). Evidence for gene transfer and expression of factor IX in haemophilia B patients treated with an AAV vector. Nat. Genet. *24*, 257–261.

Kitamura, M., Taylor, S., Unwin, R., Burton, S., Shimizu, F. and Fine, L. G. (1994). Gene transfer into the rat renal glomerulus via a mesangial cell vector: Site-specific delivery, in situ amplification, and sustained expression of an exogenous gene in vivo. J. Clin. Invest. *94*, 497–505.

Kuemmerle, N. B., Lin, P. S., Krieg, R. J., Lin, K. C., Ward, K. P. and Chan, J. C. (2000). Gene expression after intrarenal injection of plasmid DNA in the rat. Pediatr. Nephrol. *14*, 152–157.

Lai, L. W., Chan, D. M., Erickson, R. P., Hsu, S. J. and Lien, Y. H. (1998). Correction of renal tubular acidosis in carbonic anhydrase II-deficient mice with gene therapy. J. Clin. Invest. *101*, 1320–1325.

Lan, H. Y., Mu, W., Tomita, N., Huang, X. R., Li, J. H., Zhu, H. J., Morishita, R. and Johnson, R. J. (2003). Inhibition of renal fibrosis by gene transfer of inducible Smad7 using ultrasound-microbubble system in rat UUO model. J. Am. Soc. Nephrol. *14*, 1535–1548.

Lieberthal, W. and Nigam, S. K. (1998). Acute renal failure. I. Relative importance of proximal vs. distal tubular injury. Am. J. Physiol. *275 (5 Pt. 2)*, F623–F631.

Lien, Y. H. and Lai, L. W. (1997). Liposome-mediated gene transfer into the tubules. Exp. Nephrol. *5*, 132–136.

Lipkowitz, M. S., Hanss, B., Tulchin, N., Wilson, P. D., Langer, J. C., Ross, M. D., Kurtzman, G. J., Klotman, P. E. and Klotman, M. E. (1999). Transduction of renal cells in vitro and in vivo by adeno-associated virus gene therapy vectors. J. Am. Soc. Nephrol. *10*, 1908–1915.

Maruyama, H., Higuchi, N., Nishikawa, Y., Hirahara, H., Iino, N., Kameda, S., Kawachi, H., Yaoita, E., Gejyo, F. and Miyazaki, J. (2002). Kidney-targeted naked DNA transfer by retrograde renal vein injection in rats. Hum. Gene Ther. *13*, 455–468.

McCown, T. J., Xiao, S., Li, J., Breese, G. R. and Samulski, R. J. (1996). Differential and persistent expression patterns of CNS gene transfer by an adeno-associated virus (AAV) vector. Brain Res. *713*, 99–107.

McDonald, G. A., Zhu, G., Li, Y., Kovesdi, I., Wickham, T. J. and Sukhatme, V. P. (1999). Efficient adenoviral gene transfer to kidney cortical vasculature utilizing a fiber modified vector. J. Gene Med. *1*, 103–110.

Nahman, N. S., Sferra, T. J., Kronenberger, J., Urban, K. E., Troike, A. E., Johnson, A., Holycross, B. J., Nuovo, G. J. and Sedmak, D. D. (2000). Microsphere-adenoviral complexes target and transduce the glomerulus in vivo. Kidney Int. *58*, 1500–1510.

Nakao, A., Fujii, M., Matsumura, R., Kumano, K., Saito, Y., Miyazono, K. and Iwamoto, I. (1999). Transient gene transfer and expression of Smad7 prevents bleomycin-induced lung fibrosis in mice. J. Clin. Invest. *104*, 5–11.

Nakao, S., Kodama, C., Takenaka, T., Tanaka, A., Yasumoto, Y., Yoshida, A., Kanzaki, T., Enriquez, A. L., Eng, C. M., Tanaka, H., Tei, C. and Desnick, R. J. (2003). Fabry disease: Detection of undiagnosed hemodialysis patients and identification of a "renal variant" phenotype. Kidney Int. *64*, 801–807.

Nicklin, S. A., Buening, H., Dishart, K. L., de Alwis, M., Girod, A., Hacker, U., Thrasher, A. J., Ali, R. R., Hallek, M. and Baker, A. H. (2001). Efficient and selective AAV2-mediated gene transfer directed to human vascular endothelial cells. Mol. Ther. *4*, 174–181.

Noiri, E., Peresleni, T., Miller, F. and Goligorsky, M. S. (1996). In vivo targeting of inducible NO synthase with oligodeoxynucleotides protects rat kidney against ischemia. J. Clin. Invest. *97*, 2377–2383.

Park, J., Murray, G. J., Limaye, A., Quirk, J. M., Gelderman, M. P., Brady, R. O. and Qasba, P. (2003). Long-term correction of globotriaosylceramide storage in Fabry mice by recombinant adeno-associated virus-mediated gene transfer. Proc. Natl. Acad. Sci. USA *100*, 3450–3454.

Peel, A. L., Zolotukhin, S., Schrimsher, G. W., Muzyczka, N. and Reier, P. J. (1997). Efficient transduction of green fluorescent protein in spinal cord neurons using adeno-associated virus vectors containing cell type-specific promoters. Gene Ther. *4*, 16–24.

Phillips, M. I., Tang, Y., Schmidt-Ott, K., Qian, K. and Kagiyama, S. (2002). Vigilant vector: Heart-specific promoter in an adeno-associated virus vector for cardioprotection. Hypertension *39(2 Pt. 2)*, 651–655.

Pollak, M. R. (2003). The genetic basis of FSGS and steroid-resistant nephrosis. Semin. Nephrol. *23*, 141–146.

Qing, K., Mah, C., Hansen, J., Zhou, S., Dwarki, V. and Srivastava, A. (1999). Human fibroblast growth factor receptor 1 is a co-receptor for infection by adeno-associated virus 2. Nat. Med. *5*, 71–77.

Rappaport, J., Hanss, B., Kopp, J. B., Copeland, T. D., Bruggeman, L. A., Coffman, T. M. and Klotman, P. E. (1995). Transport of phosphorothioate oligonucleotides in kidney: Implications for molecular therapy. Kidney Int. *47*, 1462–1469.

Rudich, S. M., Zhou, S., Srivastava, R., Escobedo, J. A., Perez, R. V. and Manning, W. C. (2000). Dose response to a single intramuscular injection of recombinant adeno-associated virus-erythropoietin in monkeys. J. Surg. Res. *90*, 102–108.

Saito, H., Kusano, K., Kinosaki, M., Ito, H., Hirata, M., Segawa, H., Miyamoto, K. and Fukushima, N. (2003). Human fibroblast growth factor-23 mutants suppress Na+-dependent phosphate co-transport activity and 1alpha, 25-dihydroxyvitamin D3 production. J. Biol. Chem. *278*, 2206–2211.

Shayakhmetov, D. M., Carlson, C. A., Stecher, H., Li, Q., Stamatoyannopoulos, G. and Lieber, A. (2002). A high-capacity, capsid-modified hybrid adenovirus/ adeno-associated virus vector for stable transduction of human hematopoietic cells. J. Virol. 76, 1135–1143.

Snyder, R. O., Miao, C. H., Patijn, G. A., Spratt, S. K., Danos, O., Nagy, D., Gown, A. M., Winther, B., Meuse, L., Cohen, L. K., Thompson, A. R. and Kay, M. A. (1997). Persistent and therapeutic concentrations of human factor IX in mice after hepatic gene transfer of recombinant AAV vectors. Nat. Genet. 16, 270–276.

Song, S., Morgan, M., Ellis, T., Poirier, A., Chesnut, K., Wang, J., Brantly, M., Muzyczka, N., Byrne, B. J., Atkinson, M. and Flotte, T. R. (1998). Sustained secretion of human alpha-1-antitrypsin from murine muscle transduced with adeno-associated virus vectors. Proc. Natl. Acad. Sci. USA 95, 14384–14388.

Summerford, C., Bartlett, J. S. and Samulski, R. J. (1999). AlphaVbeta5 integrin: A co-receptor for adeno-associated virus type 2 infection. Nat. Med. 5, 78–82.

Summerford, C. and Samulski, R. J. (1998). Membrane-associated heparan sulfate proteoglycan is a receptor for adeno-associated virus type 2 virions. J. Virol. 72, 1438–1445.

Takahashi, H., Hirai, Y., Migita, M., Seino, Y., Fukuda, Y., Sakuraba, H., Kase, R., Kobayashi, T., Hashimoto, Y. and Shimada, T. (2002). Long-term systemic therapy of Fabry disease in a knockout mouse by adeno-associated virus-mediated muscle-directed gene transfer. Proc. Natl. Acad. Sci. USA 99, 13777–13782.

Terada, Y., Hanada, S., Nakao, A., Kuwahara, M., Sasaki, S. and Marumo, F. (2002). Gene transfer of Smad7 using electroporation of adeno virus prevents renal fibrosis in post-obstructed kidney. Kidney Int. Suppl. 61(Suppl. 1), 94–98.

Tomasoni, S., Azzollini, N., Casiraghi, F., Capogrossi, M. C., Remuzzi, G. and Benigni, A. (2000). CTLA4Ig gene transfer prolongs survival and induces donor-specific tolerance in a rat renal allograft. J. Am. Soc. Nephrol. 11, 747–752.

Tomita, N., Higaki, J., Morishita, R., Kato, K., Mikami, H., Kaneda, Y. and Ogihara, T. (1992). Direct in vivo gene introduction into rat kidney. Biochem. Biophys. Res. Commun. 186, 129–134.

Tomita, N., Morishita, R., Tomita, S., Gibbons, G. H., Zhang, L., Horiuchi, M., Kaneda, Y., Higaki, J., Ogihara, T. and Dzau, V. J. (2000). Transcription factor decoy for NFkappaB inhibits TNF-alpha-induced cytokine and adhesion molecule expression in vivo. Gene Ther. 7, 1326–1332.

Tomita, N., Morishita, R., Yamamoto, K., Higaki, J., Dzau, V. J., Ogihara, T. and Kaneda, Y. (2002). Targeted gene therapy for rat glomerulonephritis using HVJ-immunoliposomes. J. Gene Med. *4*, 527–535.

Tsujie, M., Isaka, Y., Ando, Y., Akagi, Y., Kaneda, Y., Ueda, N., Imai, E. and Hori, M. (2000). Gene transfer targeting interstitial fibroblasts by the artificial viral envelope-type hemagglutinating virus of Japan liposome method. Kidney Int. *57*, 1973–1980.

Tsujie, M., Isaka, Y., Nakamura, H., Imai, E. and Hori, M. (2001). Electroporation-mediated gene transfer that targets glomeruli. J. Am. Soc. Nephrol. *12*, 949–954.

Watnick, T. and Germino, G. G. (1999). Molecular basis of autosomal dominant polycystic kidney disease. Semin. Nephrol. *19*, 327–343.

Xiao, X., Li, J. and Samulski, R. J. (1996). Efficient long-term gene transfer into muscle tissue of immunocompetent mice by adeno-associated virus vector. J. Virol. *70*, 8098–8108.

Xu, R., Li, H., Tse, L. Y., Kung, H. F., Lu, H. and Lam, K. S. (2003). Diabetes gene therapy: Potential and challenges. Curr. Gene Ther. *3*, 65–82.

Yang, B., Ma, T., Dong, J. Y. and Verkman, A. S. (2000). Partial correction of the urinary concentrating defect in aquaporin-1 null mice by adenovirus-mediated gene delivery. Hum. Gene Ther. *11*, 567–575.

Yang, J., Dai, C. and Liu, Y. (2001). Systemic administration of naked plasmid encoding hepatocyte growth factor ameliorates chronic renal fibrosis in mice. Gene Ther. *8*, 1470–1479.

Ye, X., Jerebtsova, M., Liu, X. H., Li, Z. and Ray, P. E. (2002). Adenovirus-mediated gene transfer to renal glomeruli in rodents. Kidney Int. Suppl. *61 (Suppl. 1)*, 16–23.

Ye, X., Liu, X., Li, Z. and Ray, P. E. (2001). Efficient gene transfer to rat renal glomeruli with recombinant adenoviral vectors. Hum. Gene Ther. *12*, 141–148.

Yin, M., Wheeler, M. D., Connor, H. D., Zhong, Z., Bunzendahl, H., Dikalova, A., Samulski, R. J., Schoonhoven, R., Mason, R. P., Swenberg, J. A. and Thurman, R. G. (2001). Cu/Zn-superoxide dismutase gene attenuates ischemia-reperfusion injury in the rat kidney. J. Am. Soc. Nephrol. *12*, 2691–2700.

Yokoo, T., Ohashi, T., Utsunomiya, Y., Shen, J. S., Hisada, Y., Eto, Y., Kawamura, T. and Hosoya, T. (2001). Genetically modified bone marrow continuously supplies anti-inflammatory cells and suppresses renal injury in mouse Goodpasture syndrome. Blood *98*, 57–64.

Zhu, G., Nicolson, A. G., Cowley, B. D., Rosen, S. and Sukhatme, V. P. (1996). In vivo adenovirus-mediated gene transfer into normal and cystic rat kidneys. Gene Ther. *3*, 298–304.

CHAPTER 8

AAV for disorders of the CNS

Corinna Burger,[1,2,4] Ronald J. Mandel[2,3,4] and Nicholas Muzyczka[1,2]

[1]*Department of Molecular Genetics and Microbiology, University of Florida, Gainesville, FL 32610, USA*
[2]*Powell Gene Therapy Center, University of Florida, Gainesville, FL 32610, USA*
[3]*Department of Neuroscience, University of Florida, Gainesville, FL 32610, USA*
[4]*McKnight Brain Institute, University of Florida, Gainesville, FL 32610, USA*

8.1. Introduction

The most commonly used rAAV vector for gene therapy applications in the nervous system has been based on AAV serotype 2. rAAV2 is an efficient delivery vector to neurons in diverse anatomical regions of the nervous system (Kaplitt et al., 1994; Ali et al., 1996; McCown et al., 1996; Flannery et al., 1997; Peel et al., 1997; Bartlett et al., 1998; Klein et al., 1998; Mandel et al., 1998; Glatzel et al., 2000; Burger et al., 2004), although it also infects a small percentage of non-neuronal cell types (Kaplitt et al., 1994; McCown et al., 1996; Peel et al., 1997; Klein et al., 1998; Lo et al., 1999; Davidson et al., 2000; Cucchiarini et al., 2003). More recently, the cell tropism and transduction rate of other serotypes such as AAV1, AAV4, and AAV5, have been described in the CNS. Both rAAV1 and rAAV5 have demonstrated a larger distribution

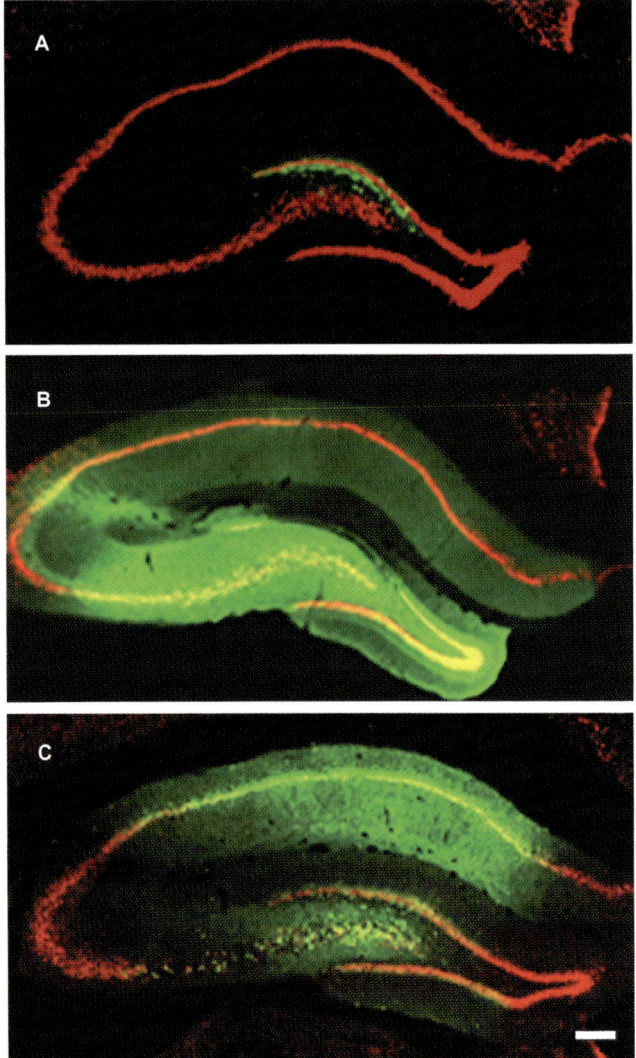

Fig. 8.1. rAAV2/1 and rAAV2/5 transduce the HPC more efficiently than rAAV2/2. (A–C), Green shows GFP fluorescence; red is NeuN immunoreactivity, which labels the neuronal cell bodies in the hippocampal formation. Yellow represents the merge of neuron-specific staining (red) and GFP (green)

and a higher number of neurons transduced than rAAV2 in the central nervous system (Alisky et al., 2000; Davidson et al., 2000; Auricchio et al., 2001; Yang et al., 2002; Passini et al., 2003; Wang et al., 2003; Weber et al., 2003; Burger et al., 2004) (Fig. 8.1). Typically the increase in volume of transduction and number of neurons transduced was five to eightfold (Burger et al., 2004). rAAV4 has been shown to target ependymal cells exclusively whereas rAAV5 transduces both neurons and ependymal cells (Davidson et al., 2000). These features of rAAV vectors make this gene delivery system extremely attractive for use in progressive disorders of the nervous system. This review describes gene therapy of neurological disorders using mainly rAAV2 as a delivery vector, however, attempts to apply other serotypes in neurological disorders are starting to emerge (Passini et al., 2003; Foust et al., 2004; Richichi et al., 2004).

8.1.1. Viral delivery into the CNS

The brain poses a challenging task for gene therapy applications. It is physically inaccessible to gene delivery, and investigators have tried different approaches to introduce therapeutic genes to the brain. These include drilling small openings in the skull for stereotaxic injection into precise brain regions, disruption of the blood–brain barrier (BBB), or injecting muscles in the hope of achieving

---

to demonstrate colocalization with rAAV transduction. All photomicrographs were taken around the injection site. (A) rAAV2/2 transduction occurs mainly in the hilus of the HPC and in the non-pyramidal neurons in the CA1 region. (B) rAAV2/5 and (C) rAAV2/1 transduce CA1–CA3 and dentate. The fluorescence intensity in (B) and (C) had to be decreased six to tenfold to match the fluorescence intensity in (A). (Scale bar = A, B, C = 200 μm). Viral titers were: rAAV2/1 = 6.2×10^{13} genome copies/ml; rAAV2/2 = 1.8×10^{13} genome copies/ml, rAAV2/5 = 1×10^{13} genome copies/ml. Two μl of vector were injected using a stereotaxic frame at a rate of 0.5 μl/min.

retrograde transport into the central nervous system (CNS). Another challenge is the distribution of transgenes globally in the brain. Focal delivery is well suited for treatment of diseases associated with regional abnormalities in the brain. But because of the global nature of diseases such as lysosomal storage disorders and most neoplastic diseases, global techniques of delivery will need to be optimized. Several approaches have been used in order to achieve systemic delivery of genes into the nervous system, including intraventricular injections or transvascular delivery (Kroll and Neuwelt, 1998; Lo et al., 1999; Davidson et al., 2000; Fu et al., 2003; Passini et al., 2003; Schlachetzki et al., 2004), multiple injections (Skorupa et al., 1999), retrograde transport from muscle (Kaspar et al., 2003), or the use of agents such as mannitol or heparin (Kroll and Neuwelt, 1998; Ghodsi et al., 1999; Nguyen et al., 2001; Georgievska et al., 2002; Mastakov et al., 2002; Fu et al., 2003; Burger et al., 2005). All of these are designed to increase the spread of virus within the brain.

Neurons are polarized cells that transport small molecules and organelles from the cell body down the axons to the synapse (anterograde transport) and back from the synapse to the cell body (retrograde transport). Several viruses have the ability to be taken up at the synapse and transported back to the cell body. These viruses include: Herpes simplex virus, adenovirus, rabies, and pseudorabies virus (Bak et al., 1977; Gillet et al., 1986; Card et al., 1990; Ghadge et al., 1995). Both anterograde and retrograde transport can be used to deliver gene products to distant places from the injection site. Reports of retrograde transport of AAV2 virus have been contradictory. Some groups have reported the lack of AAV2 retrograde transport (Chamberlin et al., 1998; Klein et al., 1999; Alisky et al., 2000; Davidson et al., 2000; Martinov et al., 2002), while others have reported that rAAV2 is capable of retrograde transport in some regions of the CNS (Kaspar et al., 2002, 2003; Boulis et al., 2003). In contrast, robust retrograde transport has been reported for AAV1 and AAV5 (Alisky et al., 2000; Burger et al., 2004; Richichi et al., 2004). Therefore, injection

Ch. 8 — AAV FOR DISORDERS OF THE CNS

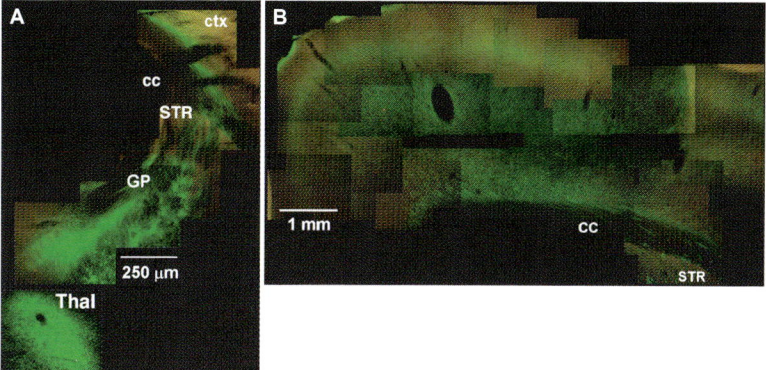

Fig. 8.2. rAAV2-UF11 transduction of the rat thalamus leads to innervation of the entire layer 4 of the cerebral cortex. Two μl of rAAV2-UF11 were injected into the anterior thalamus and the animals lived 8 weeks post-transduction. (A) The native GFP fluorescence at the transduction site in the thalamus (Thal) is shown in the lower left. Beginning dorsal to the injection site at the level of the globus pallidus (GP), a montage of the GFP+ projection through GP, then to the striatum (STR), piercing the corpus callosum (cc) and entering the cerebral cortex (ctx) is shown. (B) A lower power montage of the terminal field of the GFP+ axons are shown spreading out throughout the entire cerebral cortex ipsilateral to the thalamic injection site. This montage starts at a slightly more anterior level to the striatal section shown in (A).

of rAAV vectors into a muscle could be a potential, noninvasive approach for treating disorders such as amyotrophic lateral sclerosis (ALS). Anterograde transport of the transgene has also been reported with rAAV2 (Chamberlin et al., 1998; Kirik et al., 2000; Kaspar et al., 2002; Passini et al., 2002). This property can be exploited for global expression of a secretable transgene. For example, we have developed an approach to target genes to widespread areas of the cortex by injection into the thalamus, which contains relay nuclei that project to different cortical regions (Fig. 8.2A and B). Similarly, a single injection in the red nucleus can target a large portion of the SC (Foust et al., 2004). Finally, a few viruses travel anterogradely and transneuronally (Sun et al.,

1996; Aston-Jones and Card, 2000; Husak et al., 2000; Kelly and Strick, 2000) as virus particles. The foreign genes they express can, therefore, be used as markers for neuronal connections. This property is not shared by rAAV particles, which are completely defective for viral replication.

8.1.2. Gene therapy approaches to CNS disorders

The brain is the most complex organ in the body; therefore, there are a large number of neurological disorders affecting this system. When the gene defect is known, gene replacement is a feasible approach. However, most neurological disorders are heterogeneous, idiopathic, and/or polygenic in nature. In most neurological disorders, therefore, a single gene replacement approach is impossible. In these cases, the progression of the disorder can be ameliorated by delivery of survival agents such as neurotrophic factors. This latter approach has shown some success in disorders like Parkinson's and Alzheimer's disease. The present chapter illustrates a few examples of gene therapy strategies that have been employed in neurological disorders of the CNS using rAAV. A list of other CNS disorders that are being treated using AAV vectors is shown in Table 8.1. For a detailed review about the ongoing clinical trials using AAV for neurological disorders, please refer to Mandel and Burger (2004). The treatment of neurological disorders using other vector systems are beyond the scope of this review and the reader is directed to the following reviews: Mandel et al. (2003); Hildinger and Auricchio (2004); Schlachetzki et al. (2004); Tenenbaum et al. (2004).

8.2. Parkinson disease (PD)

PD is characterized by extensive progressive loss of the dopamine (DA) producing neurons of the substantia nigra (SN) and the concomitant loss of striatal DA terminals (McGeer et al., 1988;

TABLE 8.1
List of CNS disorders and AAV gene therapy treatments

Disorder	Gene affected	Therapeutic strategy	Delivery	References
ALS	SOD-1; unknown	Trophic factor: GDNF, IGF	Global SC: Retrograde, anterograde transport, intrathecal, intramuscular	Wang et al. (2002b); Kaspar et al. (2003); Foust et al. (2004)
Alzheimer's	APP, presenilin	Neurotrophin, vaccine	Local: Hippocampus; Global: Cortex	Zhang et al. (2003)
Batten's	CLN3P	Gene replacement	Global	Hackett et al. (2004); Rakheja et al. (2004)
Canavan's	Aspartoacylase	Gene replacement: Aspartoacylase	Global: Multiple injections, anterograde	Matalon and Michals-Matalon (1998); Leone et al. (2000); Janson et al. (2002); Matalon et al. (2003)
Epilepsy	Voltage-gated, ligand-gated ion channels; other	Antiexcitatory: Galanin, NPY, NMDA antisense	Local: Hippocampus, thalamus	Haberman et al. (2002, 2003); Lin et al. (2003); Mulley et al. (2003); Richichi et al. (2004)
Glioblastoma	N/A	Cell death	Local	Okada et al. (1996); Mizuno et al. (1998); Ma et al. (2002a,b); Yoshida et al. (2002)
Hungtington's	Huntingtin	Antiapoptotic strategies: Neurotrophins	Local: Striatum; Global cortex: Via anterograde thalamic injection	McBride et al. (2003); Foust et al. (2004)
Lysosomal storage	β-Glucuronidase	Gene replacement	Global: Multiple injections, anterograde	Skorupa et al. (1999); Daly et al. (1999a,b); Sferra et al. (2000)

(continued)

Table 8.1 (*continued*)

Disorder	Gene affected	Therapeutic strategy	Delivery	References
Parkinson's	Synuclein, parkin, UBCH1, Nurr1	Enzyme: TH, GTPCH, AADC; Growth factor: GDNF	Local: Striatum	Kang et al. (1993); Leff et al. (1999); Sanchez-Pernaute et al. (2001); Kirik et al. (2002); Muramatsu et al. (2002)
Spinal cord injury	N/A	Neurotrophin: BDNF	Global SC: Retrograde, anterograde, multiple intrathecal injections	Peel et al. (1997); Kaspar et al. (2002); Lu et al. (2003); Ruitenberg et al. (2004)

Fearnley and Lees, 1991). Typically, symptoms start to appear when about 70–80% of striatal dopamine is lost and about 50% of the dopamine neurons in the SN have degenerated. Oral administration of L-DOPA is currently the most effective and widely used medical treatment for PD.

However, as the disease progresses, patients receiving this treatment suffer from multiple side effects, most importantly wearing-off, on–off swings, and disabling dyskinesias, despite the fact that L-DOPA remains efficacious in reversing the motor symptoms of the disease. Clinical observations suggest that continuous infusion of L-DOPA (or apomorphine and long-acting DA agonists) alleviate wearing-off and dyskinetic side effects (Chase et al., 1989, 1993; Schuh and Bennett, 1993; Obeso et al., 1994). In addition, continuous, non-pulsatile stimulation of striatal DA receptors does not produce the abnormal changes in striatal gene expression that are associated with L-DOPA-induced dyskinesia (Engber et al., 1989; Doucet et al., 1996; Westin et al., 2001). Continuous L-DOPA administration instantly relieves wearing-off simply due to the static levels of L-DOPA. In addition, "on–off" motor fluctuations and peak-dose dyskinesias are also inhibited by continuous optimal-dose intravenous infusion of L-DOPA after about 9 days of treatment with a gradual decline of benefit after cessation of the continuous treatment (Schuh and Bennett, 1993; Chase et al., 1994). Based on these findings with continuous L-DOPA infusions, it has been hypothesized that L-DOPA-related complications may, at least in large part, be due to the intermittent, pulsatile supply of L-DOPA provided by the standard oral L-DOPA medication (Mouradian and Chase, 1997; Nutt, 2000). The first strategy proposed as a gene therapeutic treatment for PD was striatal L-DOPA delivery by ex vivo expression of tyrosine hydroxylase (TH) (Gage et al., 1987). The initial focus on the ex vivo delivery of TH was based on the known efficacy of peripheral L-DOPA treatment combined with the neurosurgical safety demonstrated by fetal ventral mesencephalic DA grafting in rodents and humans (Bjorklund and Lindvall, 2000).

8.2.1. Gene replacement for dopamine synthesis

In order to deliver L-DOPA, the gene for the rate-limiting enzyme for DA production, tyrosine hydroxylase must be expressed. Moreover, tetrahydrobiopterin (BH_4) is a cofactor necessary for the reaction catalyzed by TH, and the primary enzyme required for BH_4 production is GTP-cyclohydrolase 1 (GTPCH1). Given this biosynthetic pathway, one strategy for increased production of DA is vector-mediated overexpression of both TH and GTPCH1 in target tissues of the neostriatum. The advantages and disadvantages of this strategy have been reviewed in detail (Mandel et al., 1999a). We have previously reported that we could produce measurable L-DOPA levels in the 6-OHDA-lesioned rat striatum after rAAV-TH and rAAV-GTPCH1 cotransduction (Kirik et al., 2002). The data indicated that L-DOPA levels must be greater than 1.5 pmoles above the DA depleted background levels. New, high titer rAAV vectors expressing TH and GTPCH1 under the control of the chicken beta-actin hybrid promoter (Xu et al., 2001) were injected at five sites in the striatum in order to reach this level of striatal L-DOPA. Using this injection procedure in both completely DA denervated rats (MFB lesions) or partially DA depleted rats (*N. accumbens* DA innervated) revealed that when striatal L-DOPA levels exceeded the defined threshold, both partially and complete 6-OHDA lesioned rats significantly recovered both drug-induced and spontaneous motor behavior. Another important potential clinical benefit of intrastriatal rAAV-TH and rAAV-GTPCH1 delivery involves the reduction or blockade of peak dose peripheral L-DOPA-induced dyskinesias. A validated model of L-DOPA-induced dyskinesias has been recently developed for the rat (Cenci et al., 1998). rAAV-delivered intrastriatal L-DOPA in this model significantly reduced PD symptoms in the non-drug-induced spontaneous limb-use test, the cylinder test (Carlsson et al., 2005). The important conclusion of these experiments was that rAAV-mediated striatal L-DOPA expression resulted in highly significant reduction in the number of rats that developed dyskinesias by intrastriatal L-DOPA delivery. Moreover, very significantly, molecular changes that occur in the striatum in dyskinetic rats such as increased ΔFosB and prodynorphin were also reversed. In summary, at least in rodents, several key demonstrations of efficacy have been completed and marked advantages over standard L-DOPA therapy have been observed using this gene therapy approach.

Another transmitter replacement approach involves improving the effectiveness of oral L-DOPA by supplying the DA depleted neostriatum with an overabundance of the enzyme-aromatic amino-acid decarboxylase (AADC). In principle, this would convert oral L-DOPA to DA more efficiently in the striatum. In cell culture, as well as rat and monkey models of PD, rAAV-delivered AADC has been shown to increase striatal DA production in response to systemic L-DOPA administration (Kang et al., 1993; Leff et al., 1999; Shen et al., 2000; Sanchez-Pernaute et al., 2001; Muramatsu et al., 2002).

8.2.2. Neurotrophic factor delivery

Gene therapy using transmitter replacement in PD is a symptomatic therapy that does not address the underlying ongoing death of nigrostriatal DA neurons. In contrast, viral delivery of neuronal growth factors is considered a potential method to halt the ongoing progressive cell death in PD. The most potent DA-ergic neuronal growth factor studied to date is glial cell line-derived neurotrophic factor (GDNF) (Bjorklund et al., 1997). Gene therapy for PD using vector-delivered GDNF has been recently reviewed in detail (Bjorklund et al., 2000). Initially, the focus of gene transfer was the substantia nigra; however, it became clear that even though expression in the nigra protected all the DA neurons from neurotoxin-induced cell death, there was no functional impact of the GDNF treatment on animal behavior or motor recovery. This was because striatal DA innervation was critical for motor recovery and this was not preserved by GDNF expression in the nigra. In contrast, when GDNF was expressed in the striatum, striatal innervation was maintained, and substantial motor recovery was achieved (Bjorklund et al., 1997, 2000; Mandel et al., 1997). Therefore, stereotaxic injection of vector into the striatum is being seriously considered for eventual clinical trials (Bjorklund and Lindvall, 2000; Wang et al., 2002a; Eslamboli et al., 2003). For both the TH replacement discussed above and the GDNF neurotrophic strategies, the therapy relies on the fact that only a small region of the brain needs to be transduced, and the vector system can transduce the target regions in the brain completely or nearly completely. AAV2 serotype vectors transduce up to 90% of the substantia nigra in rodents, and are expected to do the same in humans. Furthermore, the newer AAV serotypes, AAV1 and AAV5, have been shown to have sufficient spread and transduction efficiency to transduce virtually all of the striatum (Burger et al., 2004). Thus, stereotaxic injection of vector into a defined region of the brain should be sufficient to achieve a therapeutic effect.

Another issue is whether it will be necessary to place either the TH or the GDNF genes under inducible transcriptional control. GDNF administration, in particular, has potential for side effects (Kordower et al., 1999; Georgievska et al., 2002; Nutt et al., 2003; Rosenblad et al., 2003). Unfortunately, the development of transcriptional regulation in the context of AAV vectors is still at an early stage and the relevant constructs need to be developed and characterized (Mansuy and Bujard, 2000; Toniatti et al., 2004) if regulation proves to be essential.

8.3. Alzheimer's disease (AD)

Alzheimer's disease is a severe neurodegenerative disorder characterized by progressive memory loss with concomitant underlying neuropathology in most cortical structures (Selkoe, 1999). The pathology is believed to be due to the accumulation of amyloid beta plaques and tangles throughout the brain. This widespread pathology is the greatest challenge for gene therapy because, under current technology, it is not feasible to transduce all of the cells in the cerebral cortex. On the other hand, if some releasable (i.e., secretable) factor is identified that can impact the underlying pathology of AD, particularly in a small region of the brain, then a gene therapeutic strategy may become more plausible for the treatment of AD.

For many years, the field of AD research has been dominated by the hypothesis that the memory deficits characteristic of this neurodegenerative disease were caused by the specific loss of a small group of cells in the basal forebrain that use acetylcholine (ACh) as their neurotransmitter, and which have widely diffuse projections to the cerebral cortex (Bartus et al., 1982). This hypothesis was strongly supported by both early neuropathological findings of reduced cortical choline acetyltransferase activity and ACh in AD, and animal model studies that showed that cholinergic depletions cause memory problems in rodents (Winkler et al., 1998). Indeed, the only currently approved drugs for treatment of memory problems in AD, function by blocking ACh metabolism in the synaptic cleft by inhibiting acetylcholinesterase (Winkler et al., 1998). However, this class of drugs is only partially effective in a subset of patients with early symptoms and the cholinergic hypothesis has been called into question (McGaughy et al., 2000). Thus, ultimately it may be necessary to find a way of removing plaques.

8.3.1. Trophic factor delivery

Nerve growth factor (NGF) (Levi-Montalcini and Angeletti, 1968), is a well established trophic factor for the cholinergic neurons of the basal forebrain (Hefti, 1986; Williams et al., 1986). Moreover, in aged rodents and primates that have been demonstrated to be deficient in cortical ACh signaling, NGF has been shown both to increase ACh synthesis and improve memory deficits (Haroutunian et al., 1986; Fischer et al., 1987, 1991; Tuszynski et al., 1991). NGF delivery as a gene therapy for AD has been tried using rAAV (Klein et al., 1999; Mandel et al., 1999b).

8.3.2. Induction of antibodies to beta amyloid

More recently, an alternative strategy has been suggested that may be useful. It has the potential for achieving a global reduction in amyloid plaques throughout the brain without the necessity of injecting vector into the brain. rAAV has been used to induce a humoral antibody response to selected transgenes (Manning et al., 1997). Indeed, relatively benign administration of vector such as oral delivery and intramuscular injection seem effective for this purpose (Xin et al., 2001, 2002). More recently, using this approach a group has proposed to induce autoantibodies to beta amyloid protein and presented proof of concept data in a transgenic animal model of Alzheimer's that suggests the approach is feasible (Zhang et al., 2003).

8.4. Epilepsy

Epilepsy is a neurological disorder of heterogeneous nature characterized by seizures that can be classified depending on the site of origin (focal versus generalized) and the frequencies of seizure episodes (seizures versus epilepsies) (Commission on classification and terminology of the International League against epilepsy, 1989; Westbrook, 2000). Seizures are caused by an imbalance in the interplay between excitation and inhibition of electrical transmission between neurons in the brain (Westbrook, 2000). Idiopathic epilepsies are the most common forms of the disorder, although the genes mutated in familial types of seizures and epilepsies are beginning to be identified, and about 25 single gene mutations have been described to date (Mulley et al., 2003; Gourfinkel-An et al., 2004).

Predictably, most of the mutations identified involve ion channels, the determinants of neuronal excitability (Mulley et al., 2003). Identification of the genetic causes of seizures has helped in understanding the mechanism of the disorder and in the discovery of new drugs that target these affected genes.

Traditional treatment of epilepsy involves anticonvulsants to block excitatory transmission. The majority of these antiepileptic drugs are known ion channel blockers [for a summary see Gourfinkel-An et al. (2004)] More drastic approaches involve surgical removal of the epileptic foci. Gene therapy approaches attempt to block excitation by blocking gene expression (Xiao et al., 1997; Haberman et al., 2002) of excitatory neurotransmitters or by overexpression of the genes involved in inhibition of excitation (Haberman et al., 2003; Lin et al., 2003; Richichi et al., 2004).

8.4.1. Galanin gene transfer

Galanin is a neuropeptide with widespread distribution in the brain (Melander et al., 1986) that has multiple neuroendocrine actions in the nervous system, including modulation of plasticity, stimulation of feeding behavior, anticonvulsant function, and antinociceptive properties (Bedecs et al., 1995). Knockout mice missing galanin or its receptor have spontaneous seizures (Mazarati et al., 2000; Jacoby et al., 2002). Conversely, overexpression of galanin has been demonstrated to have anticonvulsant properties (Mazarati et al., 1998, 2000; Kokaia et al., 2001). Therefore, this approach seems amenable to gene therapy. Two groups have tried overexpression of galanin delivered to the hilar region of the hippocampus via an AAV vector in a kainic acid-induced mouse model of seizure. Haberman et al. (2003) found that while galanin expression did not reduce the severity of the seizure, it prevented cell death of hilar neurons in the galanin-injected hemisphere when compared to the uninjected side. On the other hand, Lin et al. (2003) attributed the therapeutic effects of galanin to a reduction in the severity of seizures. The therapeutic effects reported by Lin et al. agree with previous reports where galanin was used as an anticonvulsant (Kokaia et al., 2001; Mazarati et al., 1998, 2000). Therefore, rAAV-delivered galanin appears to be a promising therapy for some forms of epilepsy. For this approach to work, regulation of expression will be necessary, because the inhibitory action of galanin in the cholinergic system is also associated with the cognitive defects observed in Alzheimer's disease (Mufson et al., 1998; Steiner et al., 2001; Counts et al., 2003).

8.4.2. Neuropeptide Y (NPY) gene transfer

NPY is another neuropeptide involved in diverse neuroendocrine functions, such as the regulation of food intake in the hypothalamus, vasoconstrictive action on the vasculature, learning and memory, antiepileptogenic function, and trophic functions [for a review see Pedrazzini et al. (2003)]. NPY mRNA levels go up after seizure, apparently as a compensatory anticonvulsant mechanism. Indeed, overexpression of NPY or one of its receptors inhibits seizures (Woldbye et al., 1996; Klapstein and Colmers, 1997; Mazarati and Wasterlain, 2002; Silva et al., 2003;). Similar to galanin, NPY has an antiepileptogenic effect by inhibiting glutamate release (Greber et al., 1994; Schwarzer et al., 1998). Richichi et al. (2004) have shown that AAV1/2 chimeras (containing both AAV1 and AAV2 capsids) overexpressing NPY in the hippocampus reduced the severity of kainic acid-induced seizures by delaying convulsion activity and by increasing the threshold to induce convulsions.

8.5. Lysosomal storage disorders (LSD)

The deficiency of any lysosomal enzyme results in accumulation of its substrate in lysosomes. Some of these diseases include Hurler syndrome, Hunter syndrome, I-cell disease, Niemann-Pick disease,

Gaucher disease, and Krabbe disease. LSDs are classified as mucopolysaccharidoses (MPS), lipidoses, or mucolipidoses depending on the nature of the stored material. Over 40 LSDs are known and they have a collective incidence of approximately 1 in 7000–8000 live births (Meikle et al., 1999; Poorthuis et al., 1999; Winchester et al., 2000). Most of the genes for these lysosomal proteins have been cloned, permitting mutation analysis in individual cases.

The MPS disorders are specifically characterized by an inherited deficiency of one of the lysosomal acid hydrolases catalyzing degradation of glycosaminoglycans (Winchester et al., 2000). Patients with MPS usually have less than 10% and often less than 1% of residual enzyme activity (Winchester et al., 2000). These enzymatic defects lead to the accumulation of mucopolysaccharides and chemically similar substrates within the lysosome (Winchester et al., 2000). Because mucopolysaccharides are found throughout the body, it is understandable that these diseases affect most organ systems (Winchester et al., 2000). The clinical symptoms of MPS include coarse faces, dysostosis multiplex, joint abnormalities, hepatosplenomegaly, corneal clouding, varying degrees of central nervous system (CNS) abnormalities, and premature death. The MPS disorders are divided into seven distinct subgroups, based on the specific enzyme deficiency (Winchester et al., 2000).

In the case of MPS VII (Sly syndrome), an extremely rare disorder, the enzyme β-glucuronidase is deficient, which leads to storage and urinary excretion of heparan sulfate and dermatan sulfate. Patients with this disorder show widely differing clinical features, ranging from severe storage disease, with coarse features, skeletal deformities, hepatosplenomegaly, and delayed development, to involvement of the aorta with dissecting aneurysm. This wide variety of clinical severity is probably due to differing genetic backgrounds of the individuals (Vogler et al., 1998).

Prospective treatments for MPS VII have been evaluated through the use of an excellent mouse model (Birkenmeier et al., 1989; Sands and Birkenmeier, 1993; Levy et al., 1996; Sands et al., 1997; Vogler et al., 1998). The MPS VII mice have clinical, phenotypic, and pathological features similar to those of human MPS VII. All of the LSDs have in common the feature that a subset of the proteins made are mannosylated and secreted into serum. The mannose 6 phosphate residues allow the protein to undergo uptake from serum via endocytosis by virgin cells. Following this, the enzyme traffics to the new lysosomes. Thus, in principle, only a small number of cells in the body need to be converted to LSD protein production to convert all cells to the β-glucuronidase positive phenotype. One difficulty with this approach, however, is that mannosylated enzyme cannot cross the blood–brain barrier (Daly et al., 1999). The combination of poor BBB penetrance of the enzyme, and the existence of an excellent mouse model, has led to a concentration of gene therapy studies in the brain of MPS VII mice. These experiments are thought of as "proof of principle" studies because there are less than 20 MPS VII patients currently alive in the US making clinical development of this strategy nearly impossible (Sands, personal communication) (Sands et al., 1997; Vogler et al., 1998). Nevertheless, MPS VII has become a proving ground for therapeutic strategies that may be useful for other LSDs.

8.5.1. β-Glucuronidase enzyme replacement

Given that the technology exists to manufacture large quantities of recombinant proteins, enzyme replacement strategies for LSDs are always a primary treatment consideration. As little as 20% of normal enzyme activity may be enough to allow normal metabolic homeostasis which bodes well for the protein replacement strategy or gene transfer. For example in Gaucher's disease, previous attempts at enzyme infusion therapy in humans have been successful. Enzyme replacement therapy has also been successful through the use of recombinant mouse β-glucuronidase with the mannose–6-P moiety (Vogler et al., 1998). β-Glucuronidase levels in injected MPS VII mouse pups were equal to or greater than those of normal mice in every tissue examined except for the brain, where 31% of normal activity was present (Vogler et al., 1998). The reduced levels in the brain are due to the fact that the enzyme has difficulty in passing the BBB. MPS VII mice treated with weekly β-glucuronidase injections from birth to 5 weeks of age had an improved phenotype and increased β-glucuronidase activity in the liver, spleen, kidney, and brain at 6 weeks of age (Sands et al., 1994).

Several investigators have used rAAV to successfully deliver β-glucuronidase to both the periphery and the brain (Daly et al., 1999a,b; Skorupa et al., 1999; Bosch et al., 2000; Sferra et al., 2000). In probably the most significant report, therapeutic levels of β-glucuronidase were achieved 1 week after neonatal IV injection of rAAV-β-glucuronidase in the liver, heart, lung, spleen, kidney, brain, and retina (Daly et al., 1999a). These levels were persistent and therapeutic for the duration of the study (16 weeks). By using neonatal mice, whose BBB integrity was still lacking, these investigators gained access to the CNS intravenously, thus avoiding a more invasive procedure later in life (Daly et al., 1999b).

Unfortunately, human therapy for LSDs will almost certainly require injection of vector into postnatal individuals. Thus, in addition to treatment of disease in peripheral organs, an effective treatment will have to include delivery to the CNS. Skorupa et al. (1999) have obtained promising data in mice that suggests this may be possible, albeit with some modifications. This group injected rAAV encoding β-glucuronidase into four locations per hemisphere (two burr holes per hemisphere and two injection per needle track). After vector injection, continuous high levels of expression were found at the sites of injection as well as secreted enzyme extending along most of the neuraxis (Skorupa et al., 1999). This resulted in widespread reversal of the pathology in the brain in the injected hemisphere that was due largely to the secretion and reuptake of β-glucuronidase through the mannose receptor. Similarly, a complete reversion of lysosomal storage granules in the enzymatically active areas, and in most of the surrounding negative areas was reported.

Fu et al. (2002) demonstrated proof of principle for a related LSD, MPS IIIB, which is deficient in the enzyme alpha-*N*-acetylglucosaminidase (NaGlu). A direct injection of 10^7 viral particles expressing the missing enzyme under the control of the NSE promoter resulted in long-term (6 months) expression of NaGlu in multiple brain structures of adult MPS IIIB mice. While vector typically transduced an area of 400–500 micron surrounding the infusion sites, the correction of glycosoaminoglycan storage involved neurons over a much broader area (1.5 mm) during the 6-month duration of experiments (Fu et al., 2002) due to secretion and reuptake of enzyme. Thus, in principle multiple injections of vector into brain parenchyma may provide significant improvement in the outcome of the neurological symptoms associated with LSDs, particularly if alternative serotypes that have a broader spread are used, or if other methods are developed to improve vector spread.

8.6. Conclusion

rAAV has proved to be an efficient and safe vector for gene delivery in the nervous system. Single injections into brain parenchyma cause little, if any, inflammatory response and only a modest humoral response to capsid antigen (Peden et al., 2004; Sanftner et al., 2004). This results in long-term gene expression that is believed to persist for the lifetime of the animal. AAV, therefore, is ideally suited for the task of gene transfer in the brain for all diseases that require long-term expression. Because vector injection in brain parenchyma results in limited diffusion (Skorupa et al., 1999; Fu et al., 2002; Burger et al., 2004), the logical candidates for rAAV gene therapy are the diseases that affect a discrete region of the brain, such as epilepsy or PD. To effectively treat diseases that affect the entire CNS (e.g., LSDs or Alzheimer's), better methods will have to be developed to distribute vector over the entire brain. In addition, many of the genes contemplated for therapy have associated toxicity when they are overexpressed. To date, reliable methods for regulating gene expression in the context of an AAV vector have not been convincingly demonstrated, and no regulated system has been tested in human trials for any disease (Mansuy and Bujard, 2000; Toniatti et al., 2004). This issue will have to be resolved as well.

Acknowledgements

Support for this work was provided by NIH Grant NS PO1 36302 to N. M. and R. J. M. N. M. is an inventor on patents related to recombinant AAV technology and owns equity in a gene therapy company that is commercializing AAV for gene therapy applications.

References

Ali, R. R. et al. (1996). Gene transfer into the mouse retina mediated by an adeno-associated viral vector. Hum. Mol. Genet. *5(5)*, 591–594.
Alisky, J. M. et al. (2000). Transduction of murine cerebellar neurons with recombinant FIV and AAV5 vectors. Neuroreport. *11(12)*, 2669–2673.
Aston-Jones, G. and Card, J. P. (2000). Use of pseudorabies virus to delineate multisynaptic circuits in brain: Opportunities and limitations. J. Neurosci. Methods *103(1)*, 51–61.
Auricchio, A. et al. (2001). Exchange of surface proteins impacts on viral vector cellular specificity and transduction characteristics: The retina as a model. Hum. Mol. Genet. *10(26)*, 3075–3081.
Bak, I. J. et al. (1977). Intraaxonal transport of Herpes simplex virus in the rat central nervous system. Brain Res. *136(3)*, 415–429.
Bartlett, J. S., Samulski, R. J. and McCown, T. J. (1998). Selective and rapid uptake of adeno-associated virus type 2 in brain. Hum. Gene Ther. *9(8)*, 1181–1186.
Bartus, R. T. et al. (1982). The cholinergic hypothesis of geriatric memory dysfunction. Science *217(4558)*, 408–414.
Bedecs, K., Berthold, M. and Bartfai, T. (1995). Galanin—10 years with a neuroendocrine peptide. Int. J. Biochem. Cell Biol. *27(4)*, 337–349.
Birkenmeier, E. H. et al. (1989). Murine mucopolysaccharidosis type VII. Characterization of a mouse with beta-glucuronidase deficiency. J. Clin. Invest. *83(4)*, 1258–1266.

Bjorklund, A. and Lindvall, O. (2000). Parkinson disease gene therapy moves toward the clinic. Nat. Med. *6(11)*, 1207–1208.

Bjorklund, A. et al. (1997). Studies on neuroprotective and regenerative effects of GDNF in a partial lesion model of Parkinson's disease. Neurobiol. Dis. *4(3–4)*, 186–200.

Bjorklund, A. et al. (2000). Towards a neuroprotective gene therapy for Parkinson's disease: Use of adenovirus, AAV and lentivirus vectors for gene transfer of GDNF to the nigrostriatal system in the rat Parkinson model. Brain Res. *886(1–2)*, 82–98.

Bosch, A. et al. (2000). Long-term and significant correction of brain lesions in adult mucopolysaccharidosis type VII mice using recombinant AAV vectors. Mol. Ther. *1(1)*, 63–70.

Boulis, N. M. et al. (2003). Intraneural colchicine inhibition of adenoviral and adeno-associated viral vector remote spinal cord gene delivery. Neurosurgery *52(2)*, 381–387.

Burger, C. et al. (2004). Recombinant AAV viral vectors pseudotyped with viral capsids from serotypes 1, 2, and 5 display differential efficiency and cell tropism after delivery to different regions of the central nervous system. Mol. Ther. *10(2)*, 302–317.

Burger, C., Nguyen, F. N., Deng, J. and Mandel, R. J. (2005). Systemic mannitol-induced hyperosmolality amplifies rAAV2 mediated striatal transduction to a greater extent than local co-infusion. Mol. Ther. *11*, 327–331.

Card, J. P. et al. (1990). Neurotropic properties of pseudorabies virus: Uptake and transneuronal passage in the rat central nervous system. J. Neurosci. *10 (6)*, 1974–1994.

Carlsson, T., Winkler, C., Burger, C., Muzyczka, N., Mandel, R. J., Cenci, M. A., Bjorklund, A. and Kirik, D. (2005). Reversal of dyskinesias in an animal model of Parkinson's disease by continuous L-DOPA delivery using recombinant AAV vectors. Brain *128*, 559–569.

Cenci, M. A., Lee, C. S. and Bjorklund, A. (1998). L-DOPA-induced dyskinesia in the rat is associated with striatal overexpression of prodynorphin- and glutamic acid decarboxylase mRNA. Eur. J. Neurosci. *10(8)*, 2694–2706.

Chamberlin, N. L. et al. (1998). Recombinant adeno-associated virus vector: Use for transgene expression and anterograde tract tracing in the CNS. Brain Res. *793(1–2)*, 169–175.

Chase, T. N., Engber, T. M. and Mouradian, M. M. (1993). Striatal dopaminoceptive system changes and motor response complications in L-dopa-treated patients with advanced Parkinson's disease. Adv. Neurol. *60*, 181–185.

Chase, T. N., Engber, T. M. and Mouradian, M. M. (1994). Palliative and prophylactic benefits of continuously administered dopaminomimetics in Parkinson's disease. Neurology *44(7 Suppl. 6)*, S15–S18.

Chase, T. N. et al. (1989). Rationale for continuous dopaminomimetic therapy of Parkinson's disease. Neurology *39(11 Suppl. 2)*, 7–10; discussion 19.

Counts, S. E. et al. (2003). Galanin in Alzheimer disease. Mol. Interv. *3(3)*, 137–156.

Cucchiarini, M. et al. (2003). Selective gene expression in brain microglia mediated via adeno-associated virus type 2 and type 5 vectors. Gene Ther. *10(8)*, 657–667.

Daly, T. M. et al. (1999a). Neonatal gene transfer leads to widespread correction of pathology in a murine model of lysosomal storage disease. Proc. Natl. Acad. Sci. USA *96(5)*, 2296–2300.

Daly, T. M. et al. (1999b). Neonatal intramuscular injection with recombinant adeno-associated virus results in prolonged beta-glucuronidase expression in situ and correction of liver pathology in mucopolysaccharidosis type VII mice. Hum. Gene Ther. *10(1)*, 85–94.

Davidson, B. L. et al. (2000). Recombinant adeno-associated virus type 2, 4, and 5 vectors: Transduction of variant cell types and regions in the mammalian central nervous system. Proc. Natl. Acad. Sci. USA *97(7)*, 3428–3432.

Doucet, J. P. et al. (1996). Chronic alterations in dopaminergic neurotransmission produce a persistent elevation of deltaFosB-like protein (s) in both the rodent and primate striatum. Eur. J. Neurosci. *8(2)*, 365–381.

Engber, T. M. et al. (1989). Continuous and intermittent levodopa differentially affect rotation induced by D-1 and D-2 dopamine agonists. Eur. J. Pharmacol. *168(3)*, 291–298.

Enquist, L. W. et al. (2002). Directional spread of an alpha-herpesvirus in the nervous system. Vet. Microbiol. *86(1–2)*, 5–16.

Eslamboli, A. et al. (2003). Recombinant adeno-associated viral vector (rAAV) delivery of GDNF provides protection against 6-OHDA lesion in the common marmoset monkey (*Callithrix jacchus*). Exp. Neurol. *184(1)*, 536–548.

Fearnley, J. M. and Lees, A. J. (1991). Ageing and Parkinson's disease: Substantia nigra regional selectivity. Brain *114(Pt. 5)*, 2283–2301.

Fischer, W. et al. (1987). Amelioration of cholinergic neuron atrophy and spatial memory impairment in aged rats by nerve growth factor. Nature *329(6134)*, 65–68.

Fischer, W. et al. (1991). NGF improves spatial memory in aged rodents as a function of age. J. Neurosci. *11(7)*, 1889–1906.

Flannery, J. G. et al. (1997). Efficient photoreceptor-targeted gene expression in vivo by recombinant adeno-associated virus. Proc. Natl. Acad. Sci. USA 94(13), 6916–6921.

Foust, K. D., Mandel, R. J., Reier, P. J. and Flotte, T. R. (2004). Anterograde delivery of gene product by rAAV5 vectors. Mol. Ther.

Fu, H. et al. (2002). Neurological correction of lysosomal storage in a mucopolysaccharidosis IIIB mouse model by adeno-associated virus-mediated gene delivery. Mol. Ther. 5(1), 42–49.

Fu, H. et al. (2003). Self-complementary adeno-associated virus serotype 2 vector: Global distribution and broad dispersion of AAV-mediated transgene expression in mouse brain. Mol. Ther. 8(6), 911–917.

Gage, F. H. et al. (1987). Grafting genetically modified cells to the brain: Possibilities for the future. Neuroscience 23(3), 795–807.

Georgievska, B., Kirik, D. and Bjorklund, A. (2002). Aberrant sprouting and downregulation of tyrosine hydroxylase in lesioned nigrostriatal dopamine neurons induced by long-lasting overexpression of glial cell line derived neurotrophic factor in the striatum by lentiviral gene transfer. Exp. Neurol. 177(2), 461–474.

Georgievska, B. et al. (2002). Neuroprotection in the rat Parkinson model by intrastriatal GDNF gene transfer using a lentiviral vector. Neuroreport 13(1), 75–82.

Ghadge, G. D. et al. (1995). CNS gene delivery by retrograde transport of recombinant replication-defective adenoviruses. Gene Ther. 2(2), 132–137.

Ghodsi, A. et al. (1999). Systemic hyperosmolality improves beta-glucuronidase distribution and pathology in murine MPS VII brain following intraventricular gene transfer. Exp. Neurol. 160(1), 109–116.

Gillet, J. P., Derer, P. and Tsiang, H. (1986). Axonal transport of rabies virus in the central nervous system of the rat. J. Neuropathol. Exp. Neurol. 45 (6), 619–634.

Glatzel, M. et al. (2000). Adenoviral and adeno-associated viral transfer of genes to the peripheral nervous system. Proc. Natl. Acad. Sci. USA 97(1), 442–447.

Gourfinkel-An, I. et al. (2004). Monogenic idiopathic epilepsies. Lancet Neurol. 3(4), 209–218.

Greber, S., Schwarzer, C. and Sperk, G. (1994). Neuropeptide Y inhibits potassium-stimulated glutamate release through Y2 receptors in rat hippocampal slices in vitro. Br. J. Pharmacol. 113(3), 737–740.

Haberman, R. et al. (2002). Therapeutic liabilities of in vivo viral vector tropism: Adeno-associated virus vectors, NMDAR1 antisense, and focal seizure sensitivity. Mol. Ther. 6(4), 495–500.

Haberman, R. P., Samulski, R. J. and McCown, T. J. (2003). Attenuation of seizures and neuronal death by adeno-associated virus vector galanin expression and secretion. Nat. Med. *9(8)*, 1076–1080.

Hackett, N. R., Redmond, Jr.,E. D., Sondhi, D., Giannaris, E. L., Vassallo, E., Kaminsky, S. M. and Crystal, R. G. (2004). Safety of administration of AAV2CUhCLN2, a candidate treatment for late infantile neuronal lipofuscinosis to the brain of rats and non-human primates. Mol. Ther. *9(Suppl. 1)*, s165.

Haroutunian, V., Kanof, P. D. and Davis, K. L. (1986). Partial reversal of lesion-induced deficits in cortical cholinergic markers by nerve growth factor. Brain Res. *386(1–2)*, 397–399.

Hefti, F. (1986). Nerve growth factor promotes survival of septal cholinergic neurons after fimbrial transactions. J. Neurosci. *6(8)*, 2155–2162.

Hildinger, M. and Auricchio, A. (2004). Advances in AAV-mediated gene transfer for the treatment of inherited disorders. Eur. J. Hum. Genet. *12 (4)*, 263–271.

Husak, P. J., Kuo, T. and Enquist, L. W. (2000). Pseudorabies virus membrane proteins gI and gE facilitate anterograde spread of infection in projection-specific neurons in the rat. J. Virol. *74(23)*, 10975–10983.

Jacoby, A. S. et al. (2002). Critical role for GALR1 galanin receptor in galanin regulation of neuroendocrine function and seizure activity. Brain Res. Mol. Brain Res. *107(2)*, 195–200.

Janson, C. et al. (2002). Clinical protocol Gene therapy of Canavan disease: AAV–2 vector for neurosurgical delivery of aspartoacylase gene (ASPA) to the human brain. Hum. Gene Ther. *13(11)*, 1391–1412.

Kang, U. J. et al. (1993). Regulation of dopamine production by genetically modified primary fibroblasts. J. Neurosci. *13(12)*, 5203–5211.

Kaplitt, M. G. et al. (1994). Long-term gene expression and phenotypic correction using adeno-associated virus vectors in the mammalian brain. Nat. Genet. *8(2)*, 148–154.

Kaspar, B. K. et al. (2002). Targeted retrograde gene delivery for neuronal protection. Mol. Ther. *5(1)*, 50–56.

Kaspar, B. K. et al. (2003). Retrograde viral delivery of IGF–1 prolongs survival in a mouse ALS model. Science *301(5634)*, 839–842.

Kelly, R. M. and Strick, P. L. (2000). Rabies as a transneuronal tracer of circuits in the central nervous system. J. Neurosci. Methods *103(1)*, 63–71.

Kirik, D. et al. (2000). Long-term rAAV-mediated gene transfer of GDNF in the rat Parkinson's model: Intrastriatal but not intranigral transduction promotes functional regeneration in the lesioned nigrostriatal system. J. Neurosci. *20(12)*, 4686–4700.

Kirik, D. et al. (2002). Reversal of motor impairments in parkinsonian rats by continuous intrastriatal delivery of L-dopa using rAAV-mediated gene transfer. Proc. Natl. Acad. Sci. USA 99(7), 4708–4713.
Klapstein, G. J. and Colmers, W. F. (1997). Neuropeptide Y suppresses epileptiform activity in rat hippocampus in vitro. J. Neurophysiol. 78(3), 1651–1661.
Klein, R. L. et al. (1998). Neuron-specific transduction in the rat septohippocampal or nigrostriatal pathway by recombinant adeno-associated virus vectors. Exp. Neurol. 150(2), 183–194.
Klein, R. L. et al. (1999). Long-term actions of vector-derived nerve growth factor or brain-derived neurotrophic factor on choline acetyltransferase and Trk receptor levels in the adult rat basal forebrain. Neuroscience 90 (3), 815–821.
Kokaia, M. et al. (2001). Suppressed kindling epileptogenesis in mice with ectopic overexpression of galanin. Proc. Natl. Acad. Sci. USA 98(24), 14006–14011.
Kordower, J. H. et al. (1999). Clinicopathological findings following intraventricular glial-derived neurotrophic factor treatment in a patient with Parkinson's disease. Ann. Neurol. 46(3), 419–424.
Kroll, R. A. and Neuwelt, E. A. (1998). Outwitting the blood–brain barrier for therapeutic purposes: Osmotic opening and other means. Neurosurgery 42 (5), 1083–1099 (discussion 1099–1100).
Leff, S. E. et al. (1999). Long-term restoration of striatal L-aromatic amino acid decarboxylase activity using recombinant adeno-associated viral vector gene transfer in a rodent model of Parkinson's disease. Neuroscience 92 (1), 185–196.
Leone, P. et al. (2000). Aspartoacylase gene transfer to the mammalian central nervous system with therapeutic implications for Canavan disease. Ann. Neurol. 48(1), 27–38.
Levi-Montalcini, R. and Angeletti, P. U. (1968). Nerve growth factor. Physiol. Rev. 48(3), 534–569.
Levy, B. et al. (1996). Neuropathology of murine mucopolysaccharidosis type VII. Acta Neuropathol. (Berl.) 92(6), 562–568.
Lin, E. J. et al. (2003). Recombinant AAV-mediated expression of galanin in rat hippocampus suppresses seizure development. Eur. J. Neurosci. 18(7), 2087–2092.
Lo, W. D. et al. (1999). Adeno-associated virus-mediated gene transfer to the brain: Duration and modulation of expression. Hum. Gene Ther. 10(2), 201–213.
Lu, Y. Y. et al. (2003). Intramuscular injection of AAV-GDNF results in sustained expression of transgenic GDNF, and its delivery to spinal motoneurons by retrograde transport. Neurosci. Res. 45(1), 33–40.

Ma, H. I. et al. (2002a). Intratumoral gene therapy of malignant brain tumor in a rat model with angiostatin delivered by adeno-associated viral (AAV) vector. Gene Ther. *9(1)*, 2–11.

Ma, H. I. et al. (2002b). Suppression of intracranial human glioma growth after intramuscular administration of an adeno-associated viral vector expressing angiostatin. Cancer Res. *62(3)*, 756–763.

Mandel, R. J., Velardo, J., Sullivan, S. M., Rodriguez, E., Piercefield, E., Deng, J., Socarras, C., Reier, P. and Burger, C. (2003). Gene therapy for disorders of the central nervous system. In: Textbook of Neurointensive Care (Layon, A. J., Gabrielli, A. and Friedman, W., eds.). Saunders, pp. 865–890.

Mandel, R. J. and Burger, C. (2004). Clinical trials in neurological disorders using AAV vectors: Promises and challenges. Curr. Op. Mol. Therapeutics *6(5)*, 482–490.

Mandel, R. J. et al. (1997). Midbrain injection of recombinant adeno-associated virus encoding rat glial cell line-derived neurotrophic factor protects nigral neurons in a progressive 6-hydroxydopamine-induced degeneration model of Parkinson's disease in rats. Proc. Natl. Acad. Sci. USA *94(25)*, 14083–14088.

Mandel, R. J. et al. (1998). Characterization of intrastriatal recombinant adeno-associated virus-mediated gene transfer of human tyrosine hydroxylase and human GTP-cyclohydrolase I in a rat model of Parkinson's disease. J. Neurosci. *18(11)*, 4271–4284.

Mandel, R. J. et al. (1999a). Progress in direct striatal delivery of L-dopa via gene therapy for treatment of Parkinson's disease using recombinant adeno-associated viral vectors. Exp. Neurol. *159(1)*, 47–64.

Mandel, R. J. et al. (1999b). Nerve growth factor expressed in the medial septum following in vivo gene delivery using a recombinant adeno-associated viral vector protects cholinergic neurons from fimbria-fornix lesion-induced degeneration. Exp. Neurol. *155(1)*, 59–64.

Manning, W. C. et al. (1997). Genetic immunization with adeno-associated virus vectors expressing herpes simplex virus type 2 glycoproteins B and D. J. Virol. *71(10)*, 7960–7962.

Mansuy, I. M. and Bujard, H. (2000). Tetracycline-regulated gene expression in the brain. Curr. Opin. Neurobiol. *10(5)*, 593–596.

Martinov, V. N. et al. (2002). Targeting functional subtypes of spinal motoneurons and skeletal muscle fibers in vivo by intramuscular injection of adenoviral and adeno-associated viral vectors. Anat. Embryol. (Berl.) *205(3)*, 215–221.

Mastakov, M. Y. et al. (2002). Recombinant adeno-associated virus serotypes 2- and 5-mediated gene transfer in the mammalian brain: Quantitative analysis of heparin co-infusion. Mol. Ther. *5(4)*, 371–380.

Matalon, R. and Michals-Matalon, K. (1998). Molecular basis of Canavan disease. Eur. J. Paediatr. Neurol. *2(2)*, 69–76.

Matalon, R. et al. (2003). Adeno-associated virus-mediated aspartoacylase gene transfer to the brain of knockout mouse for canavan disease. Mol. Ther. *7(5 Pt. 1)*, 580–587.

Mazarati, A. and Wasterlain, C. G. (2002). Anticonvulsant effects of four neuropeptides in the rat hippocampus during self-sustaining status epilepticus. Neurosci. Lett. *331(2)*, 123–127.

Mazarati, A. M. et al. (1998). Galanin modulation of seizures and seizure modulation of hippocampal galanin in animal models of status epilepticus. J. Neurosci. *18(23)*, 10070–10077.

Mazarati, A. M. et al. (2000). Modulation of hippocampal excitability and seizures by galanin. J. Neurosci. *20(16)*, 6276–6281.

McBride, J. L. et al. (2003). Structural and functional neuroprotection in a rat model of Huntington's disease by viral gene transfer of GDNF. Exp. Neurol. *181(2)*, 213–223.

McCown, T. J. et al. (1996). Differential and persistent expression patterns of CNS gene transfer by an adeno-associated virus (AAV) vector. Brain Res. *713(1–2)*, 99–107.

McGaughy, J. et al. (2000). The role of cortical cholinergic afferent projections in cognition: Impact of new selective immunotoxins. Behav. Brain Res. *115 (2)*, 251–263.

McGeer, P. L. et al. (1988). Rate of cell death in parkinsonism indicates active neuropathological process. Ann. Neurol. *24(4)*, 574–576.

Meikle, P. J. et al. (1999). Prevalence of lysosomal storage disorders. JAMA *281(3)*, 249–254.

Melander, T., Hokfelt, T. and Rokaeus, A. (1986). Distribution of galaninlike immunoreactivity in the rat central nervous system. J. Comp. Neurol. *248 (4)*, 475–517.

Mizuno, M. et al. (1998). Adeno-associated virus vector containing the herpes simplex virus thymidine kinase gene causes complete regression of intracerebrally implanted human gliomas in mice, in conjunction with ganciclovir administration. Jpn. J. Cancer Res. *89(1)*, 76–80.

Mouradian, M. M. and Chase, T. N. (1997). Gene therapy for Parkinson's disease: An approach to the prevention or palliation of levodopa-associated motor complications. Exp. Neurol. *144(1)*, 51–57.

Mufson, E. J. et al. (1998). Galanin expression within the basal forebrain in Alzheimer's disease Comments on therapeutic potential. Ann. N.Y. Acad. Sci. *863*, 291–304.

Mulley, J. C. et al. (2003). Channelopathies as a genetic cause of epilepsy. Curr. Opin. Neurol. *16(2)*, 171–176.

Muramatsu, S. et al. (2002). Behavioral recovery in a primate model of Parkinson's disease by triple transduction of striatal cells with adeno-associated viral vectors expressing dopamine-synthesizing enzymes. Hum. Gene Ther. *13(3)*, 345–354.

Nguyen, J. B. et al. (2001). Convection-enhanced delivery of AAV-2 combined with heparin increases TK gene transfer in the rat brain. Neuroreport *12(9)*, 1961–1964.

Nutt, J. G. (2000). Clinical pharmacology of levodopa-induced dyskinesia. Ann. Neurol. *47(4 Suppl. 1)*, S160–S164 (discussion S164–S166).

Nutt, J. G. et al. (2003). Randomized, double-blind trial of glial cell line-derived neurotrophic factor (GDNF) in PD. Neurology *60(1)*, 69–73.

Obeso, J. A. et al. (1994). The role of pulsatile versus continuous dopamine receptor stimulation for functional recovery in Parkinson's disease. Eur. J. Neurosci. *6(6)*, 889–897.

Okada, H. et al. (1996). Gene therapy against an experimental glioma using adeno-associated virus vectors. Gene Ther. *3(11)*, 957–964.

Passini, M. A. et al. (2002). Distribution of a lysosomal enzyme in the adult brain by axonal transport and by cells of the rostral migratory stream. J. Neurosci. *22(15)*, 6437–6446.

Passini, M. A. et al. (2003). Intraventricular brain injection of adeno-associated virus type 1 (AAV1) in neonatal mice results in complementary patterns of neuronal transduction to AAV2 and total long-term correction of storage lesions in the brains of beta-glucuronidase-deficient mice. J. Virol. *77(12)*, 7034–7040.

Peden, C. S. et al. (2004). Circulating anti-wild-type adeno-associated virus type 2 (AAV2) antibodies inhibit recombinant AAV2 (rAAV2)-mediated, but not rAAV5-mediated, gene transfer in the brain. J. Virol. *78(12)*, 6344–6359.

Pedrazzini, T., Pralong, F. and Grouzmann, E. (2003). Neuropeptide Y: The universal soldier. Cell. Mol. Life Sci. *60(2)*, 350–377.

Peel, A. L. et al. (1997). Efficient transduction of green fluorescent protein in spinal cord neurons using adeno-associated virus vectors containing cell type-specific promoters. Gene Ther. *4(1)*, 16–24.

Poorthuis, B. J. et al. (1999). The frequency of lysosomal storage diseases in the Netherlands. Hum. Genet. *105(1–2)*, 151–156.

Proposal for revised classification of epilepsies and epileptic syndromes (1989). Commission on classification and terminology of the International League against epilepsy. *Epilepsia* **30** 389–399.

Rakheja, D. et al. (2004). CLN3P, the Batten disease protein, localizes to membrane lipid rafts (detergent-resistant membranes). Biochem. Biophys. Res. Commun. *317(4)*, 988–991.

Richichi, C. et al. (2004). Anticonvulsant and antiepileptogenic effects mediated by adeno-associated virus vector neuropeptide y expression in the rat hippocampus. J. Neurosci. *24(12)*, 3051–3059.

Rosenblad, C., Georgievska, B. and Kirik, D. (2003). Long-term striatal overexpression of GDNF selectively downregulates tyrosine hydroxylase in the intact nigrostriatal dopamine system. Eur. J. Neurosci. *17(2)*, 260–270.

Ruitenberg, M. J. et al. (2004). Adeno-associated viral vector-mediated gene transfer of brain-derived neurotrophic factor reverses atrophy of rubrospinal neurons following both acute and chronic spinal cord injury. Neurobiol. Dis. *15(2)*, 394–406.

Sanchez-Pernaute, R. et al. (2001). Functional effect of adeno-associated virus mediated gene transfer of aromatic L-amino acid decarboxylase into the striatum of 6-OHDA-lesioned rats. Mol. Ther. *4(4)*, 324–330.

Sands, M. S. and Birkenmeier, E. H. (1993). A single-base-pair deletion in the beta-glucuronidase gene accounts for the phenotype of murine mucopolysaccharidosis type VII. Proc. Natl. Acad. Sci. USA *90(14)*, 6567–6571.

Sands, M. S. et al. (1994). Enzyme replacement therapy for murine mucopolysaccharidosis type VII. J. Clin. Invest. *93(6)*, 2324–2331.

Sands, M. S. et al. (1997). Gene therapy for murine mucopolysaccharidosis type VII. Neuromuscul. Disord. *7(5)*, 352–360.

Sanftner, L. M. et al. (2004). , Striatal delivery of rAAV-hAADC to rats with preexisting immunity to AAV. Mol. Ther. *9(3)*, 403–409.

Schlachetzki, F. et al. (2004). Gene therapy of the brain: The trans-vascular approach. Neurology *62(8)*, 1275–1281.

Schuh, L. A. and Bennett, Jr., J. P. (1993). Suppression of dyskinesias in advanced Parkinson's disease I. Continuous intravenous levodopa shifts dose response for production of dyskinesias but not for relief of parkinsonism in patients with advanced Parkinson's disease. Neurology *43(8)*, 1545–1550.

Schwarzer, C., Kofler, N. and Sperk, G. (1998). Up-regulation of neuropeptide Y-Y2 receptors in an animal model of temporal lobe epilepsy. Mol. Pharmacol. *53(1)*, 6–13.

Selkoe, D. J. (1999). Translating cell biology into therapeutic advances in Alzheimer's disease. Nature *399(6738 Suppl.)*, A23–A31.

Sferra, T. J. et al. (2000). Recombinant adeno-associated virus-mediated correction of lysosomal storage within the central nervous system of the adult mucopolysaccharidosis type VII mouse. Hum. Gene Ther. *11(4)*, 507–519.

Shen, Y. et al. (2000). Triple transduction with adeno-associated virus vectors expressing tyrosine hydroxylase, aromatic-L-amino-acid decarboxylase, and GTP cyclohydrolase I for gene therapy of Parkinson's disease. Hum. Gene Ther. *11(11)*, 1509–1519.

Silva, A. P. et al. (2003). Activation of neuropeptide Y receptors is neuroprotective against excitotoxicity in organotypic hippocampal slice cultures. FASEB J. *17(9)*, 1118–1120.

Skorupa, A. F. et al. (1999). Sustained production of beta-glucuronidase from localized sites after AAV vector gene transfer results in widespread distribution of enzyme and reversal of lysosomal storage lesions in a large volume of brain in mucopolysaccharidosis VII mice. Exp. Neurol. *160(1)*, 17–27.

Steiner, R. A. et al. (2001). Galanin transgenic mice display cognitive and neurochemical deficits characteristic of Alzheimer's disease. Proc. Natl. Acad. Sci. USA *98(7)*, 4184–4189.

Sun, N., Cassell, M. D. and Perlman, S. (1996). Anterograde, transneuronal transport of herpes simplex virus type 1 strain H129 in the murine visual system. J. Virol. *70(8)*, 5405–5413.

Tenenbaum, L. et al. (2004). Recombinant AAV-mediated gene delivery to the central nervous system. J. Gene Med. *6(Suppl. 1)*, S212–S222.

Toniatti, C. et al. (2004). Gene therapy progress and prospects: Transcription regulatory systems. Gene Ther. *11(8)*, 649–657.

Tuszynski, M. H. et al. (1991). Recombinant human nerve growth factor infusions prevent cholinergic neuronal degeneration in the adult primate brain. Ann. Neurol. *30(5)*, 625–636.

Vogler, C. et al. (1998). Murine mucopolysaccharidosis type VII: The impact of therapies on the clinical course and pathology in a murine model of lysosomal storage disease. J. Inherit. Metab. Dis. *21(5)*, 575–586.

Wang, C. et al. (2003). Recombinant AAV serotype 1 transduction efficiency and tropism in the murine brain. Gene Ther. *10(17)*, 1528–1534.

Wang, L. J. et al. (2002a). Delayed delivery of AAV-GDNF prevents nigral neurodegeneration and promotes functional recovery in a rat model of Parkinson's disease. Gene Ther. *9(6)*, 381–389.

Wang, L. J. et al. (2002b). Neuroprotective effects of glial cell line-derived neurotrophic factor mediated by an adeno-associated virus vector in a transgenic animal model of amyotrophic lateral sclerosis. J. Neurosci. *22 (16)*, 6920–6928.

Weber, M. et al. (2003). Recombinant adeno-associated virus serotype 4 mediates unique and exclusive long-term transduction of retinal pigmented epithelium in rat, dog, and nonhuman primate after subretinal delivery. Mol. Ther. *7(6)*, 774–781.

Westbrook, G. (2000). Seizures and epilepsy. In: Principles of Neural Science (Kandel, E. R., Schwartz, J. H. and Jessell, T. M., eds.). pp. 910–935.

Westin, J. E. et al. (2001). Persistent changes in striatal gene expression induced by long-term L-DOPA treatment in a rat model of Parkinson's disease. Eur. J. Neurosci. *14(7)*, 1171–1176.

Williams, L. R. et al. (1986). Continuous infusion of nerve growth factor prevents basal forebrain neuronal death after fimbria fornix transection. Proc. Natl. Acad. Sci. USA *83(23)*, 9231–9235.

Winchester, B., Vellodi, A. and Young, E. (2000). The molecular basis of lysosomal storage diseases and their treatment. Biochem. Soc. Trans. *28 (2)*, 150–154.

Winkler, J. et al. (1998). Cholinergic strategies for Alzheimer's disease. J. Mol. Med. *76(8)*, 555–567.

Woldbye, D. P. et al. (1996). Neuropeptide Y inhibits hippocampal seizures and wet dog shakes. Brain Res. *737(1–2)*, 162–168.

Xiao, X. et al. (1997). Adeno-associated virus (AAV) vector antisense gene transfer in vivo decreases GABA(A) alpha1 containing receptors and increases inferior collicular seizure sensitivity. Brain Res. *756(1–2)*, 76–83.

Xin, K. Q. et al. (2001). A novel recombinant adeno-associated virus vaccine induces a long-term humoral immune response to human immunodeficiency virus. Hum. Gene Ther. *12(9)*, 1047–1061.

Xin, K. Q. et al. (2002). Oral administration of recombinant adeno-associated virus elicits human immunodeficiency virus-specific immune responses. Hum. Gene Ther. *13(13)*, 1571–1581.

Xu, L. et al. (2001). CMV-beta-actin promoter directs higher expression from an adeno-associated viral vector in the liver than the cytomegalovirus or elongation factor 1 alpha promoter and results in therapeutic levels of human factor X in mice. Hum. Gene Ther. *12(5)*, 563–573.

Yang, G. S. et al. (2002). Virus-mediated transduction of murine retina with adeno-associated virus: Effects of viral capsid and genome size. J. Virol. *76 (15)*, 7651–7660.

Yoshida, J. et al. (2002). Antitumor effect of an adeno-associated virus vector containing the human interferon-beta gene on experimental intracranial human glioma. Jpn. J. Cancer Res. *93(2)*, 223–228.

Zhang, J. et al. (2003). A novel recombinant adeno-associated virus vaccine reduces behavioral impairment and beta-amyloid plaques in a mouse model of Alzheimer's disease. Neurobiol. Dis. *14(3)*, 365–379.

Laboratory Techniques in Biochemistry and Molecular Biology, Volume 31
Adeno-Associated Viral Vectors for Gene Therapy
T. R. Flotte and K. I. Berns (Editors)

CHAPTER 9

Gene therapy for cardiovascular applications

Christina A. Pacak,[1,2] Cathryn Mah[1,2,3] and Barry J. Byrne[1,2,3]

[1]Department of Pediatrics, [2]Department of Molecular Genetics and Microbiology, [3]Division of Cell and Molecular Therapies, University of Florida, Box 100296, Gainesville, FL 32610-0296, USA

In the US, cardiovascular disease accounted for 38.5% of total deaths in the year 2001 and overall, 20% of Americans have some form of cardiovascular disease (American Heart Assoc., 2004). While the accumulated disorders that make up the broad category of cardiovascular disease are often influenced by individual lifestyle choices, they are also frequently the consequence of an inherited genetic predisposition or a congenital metabolic defect. Among these are such common cardiovascular problems as myocardial infarction, atherosclerosis, stroke, and dilated cardiomyopathy, in addition to relatively rare problems such as ion channel abnormalities, structural protein defects, or enzymatic deficiency. Many of these diseases are currently managed using pharmaceutical or surgical approaches whenever possible but can potentially be treated with gene therapy some day. Ideally, a single administration of a gene delivery vehicle containing the missing, or mutated gene of interest will be sufficient to provide persistent expression throughout the required amount of time in specifically targeted tissues of an afflicted individual (gain of function). Strategies are likewise emerging for reversal of a dominant negative gene effect (loss of function).

Several questions must be addressed when developing a gene therapy strategy for treatment of a specific problem. The first question is what is the necessary duration of expression of the delivered product in order to achieve a therapeutic effect? This obviously depends greatly on the problem in hand. Is it a temporary therapy to aid in wound healing, promote therapeutic angiogenesis, or treat a transitory ailment? Or, is it a lifelong defect as a result of either an inherited or acquired disease requiring sustained transgene expression? Once the duration of expression is determined, the minimum target-for-treatment must be established. If the delivered product is a secreted protein, perhaps targeting a small number of cells will suffice as a provisional depot that can provide for the entire body. However, when attempting delivery of a transgene encoding a membrane-bound protein, only those cells that are successfully transduced by the delivery system will display expression of the transgene. Good examples of the latter concept are the gene therapy approaches for treatment of the muscular dystrophies. In such cases, there may be a threshold minimum percentage of cells necessary to be corrected in order to observe functional correction and impede disease progression. Another important consideration when developing a gene therapy strategy is determining whether the treatment should be delivered locally for expression in a distinct area, or systemically in order to achieve a global distribution of the transgene product. The gene delivery systems currently being studied can be divided into two major categories: Viral and non-viral based. Viral-based gene delivery systems include retrovirus, adenovirus (Ad), lentivirus, and adeno-associated virus (AAV). The aim of this chapter is to provide a general overview of the progress made in the gene therapy field to provide the necessary amount of gene expression in various models of cardiovascular disease in order to demonstrate both functional and morphological correction in the heart.

9.1. Viral gene delivery systems

9.1.1. Retroviruses

The initial viral-based gene delivery experiments were performed in vitro using replication defective retroviral vectors in an attempt to transduce endothelial cells (Yao et al., 1991). Preliminary viral-based in vivo cardiovascular gene delivery experiments were also performed using retroviral vectors and demonstrated this vector system's ability to successfully transduce vascular endothelial cells (Newman et al., 1991). Although attempts at direct transduction of the carotid artery were not successful, it was established that efficient endothelial transduction could be achieved through the retrovirus-mediated infection of cells ex vivo and their subsequent arterial seeding in vivo (Lynch et al., 1992).

Other studies have been aimed at demonstrating transduction of the injured myocardium following myocardial infarction. Investigators have found that successful transgene expression could be achieved following injection of the TE-FLY-A-based MFG retroviral vector into the injured region of a beating rat heart after local freeze-thaw injury, although transduction efficiency was only 14% of the local myofibroblasts (Byun et al., 2000). Several gene transfer restrictions with the retroviral delivery system include the vector's inability to transduce nonproliferating target cells (Psarras et al., 2004) and the potential for insertional mutagenesis and activation of oncogenes following host chromosome integration (Haviernik and Bunting, 2004).

9.1.2. Adenoviruses

Another viral-based gene delivery system that has been extensively studied is the adenovirus (Ad) vector. One of the first in vivo Ad-mediated gene transfer experiments was performed with the use of a recombinant vector expressing a β-galactosidase reporter gene. Mouse neonates were intravenously (IV) injected with the Ad-β-galactosidase vector. Transgene expression was observed in heart, skeletal muscle, lung, liver, and intestine for 15 days following vector administration (Stratfordperricaudet et al., 1992). Direct injections of Ad-β-galactosidase into the myocardium using the sub-diaphragmatic approach also resulted in successful cardiomyocyte transduction with peak enzymatic activities during the first week following administration (Guzman et al., 1993). Recently, Walsh et al. were able to demonstrate the ability of in vivo Ad-mediated Akt proto-oncogene delivery to limit infarct size when administered to rat hearts following ischemia-reperfusion injury (Miao et al., 2000). Although one of the benefits of using this delivery system is that onset of Ad-mediated transgene expression has been observed as early as 6 h post-injection (Roks et al., 1997), one of the limitations is the relatively short duration of expression of 2–3 weeks (Lemarchand et al., 1993) and the potential for the elicitation of an immune response (O'Donnell et al., 2001).

9.1.3. Lentiviruses

Lentivirus-based vectors have also been utilized for cardiovascular gene delivery. Preliminary experiments involved the use of lentivirus for the in vitro delivery of genes into both dividing and non-dividing cells in tissue culture (Frimpong and Spector, 2000). The promising results of these studies led researchers to attempt ex vivo gene transfer into cells followed by vascular seeding much like what had been previously demonstrated with retrovirus gene delivery. Direct cardiac gene transfer was performed using an HIV-1-based lentivirus carrying the enhanced green fluorescent protein (E-GFP) gene and was found to successfully transduce adult mouse cardiomyocytes as efficiently as Ad-based vectors (Zhao et al., 2002). In addition, prolonged, extensive β-galactosidase expression was observed in heart as well as liver, brain, and muscle following administration of a lentivirus-based vector to the fetal circulation of immuno-competent (MF1 mice) (Waddington et al., 2003). This proof of principal experiment demonstrated the ability for efficient prenatal gene delivery and its potential for permanent disease correction (Waddington et al., 2003).

More recently, studies have shown that the generation of lentiviral vectors, pseudotyped with different viral glycoproteins, may allow for manipulation of natural tropisms in order to target specific tissues (Verhoeyen and Cosset, 2004). Some of the limitations of the lentivirus vector system include the inability to transduce progenitor hematopoietic stem cells in G(0), non-activated primary blood lymphocytes, or monocytes (Verhoeyen and Cosset, 2004). In addition, the random insertion of lentivirus genomes into the host genome increases the possibility for transgene expression to be affected by flanking host chromatin (Yee and Zaia, 2004).

9.1.4. Adeno-associated virus

Adeno-associated virus AAV has emerged as an attractive viral-based choice for cardiovascular gene therapy due to its small size, safety, and proven ability to persist for long periods of time in skeletal muscle. Initially, cardiovascular gene delivery experiments using recombinant AAV serotype 2 were performed using striated muscle (Kessler et al., 1996) and a rat carotid artery model (Gnatenko et al., 1997). Recombinant AAV (rAAV) carrying the β-galactosidase marker gene successfully transduced endothelial and vascular smooth muscle cells with frequencies approaching 90%, with no evidence of disruption of the vessel architecture (Gnatenko et al., 1997). More recently, cardiac gene delivery was demonstrated using a method of direct infusion of an rAAV2 vector encoding for a therapeutic transgene into the coronary artery ex vivo in a heterotopically transplanted heart (Asfour et al., 2002; Li et al., 2003). Effective gene transfer was achieved in up to 90% of the cardiomyocytes and in a corresponding marker–gene experiment, persisted for more than 1 year without immune rejection (Li et al., 2003).

Through the recent discovery of new serotypes of AAV and capsid manipulations of those serotypes, researchers have made it possible to alter natural tropisms in order to target and/or evade specific organs (Chao et al., 2000; Hauck et al., 2003; Nicklin et al., 2003; Du et al., 2004; Warrington et al., 2004). This is essential because depending upon the specific disease being treated, applications in which either gene transfer into cardiac tissue alone or global transduction throughout all cardiac and skeletal muscle, may be necessary for disease correction. One of the major limitations to AAV-mediated gene delivery is transgene size. This challenge has recently been overcome by investigators studying Duchenne's muscular dystrophy through the development of the mini-dystrophin gene that contains only those sequences that are essential for the translation of a functional dystrophin protein. Direct injections of rAAV carrying the mini-dystrophin gene into mouse neonate chest cavities has been shown to successfully restore the dystrophin–glycoprotein complex in cardiac tissue up to 10 months post-administration (Yue et al., 2003). Using the cardiomyopathic hamster model, investigators have shown that rAAV-mediated cardiac gene delivery of a pseudophosphorylated mutant of phospholamban was successfully able to enhance myocardial uptake of calcium and suppress impairment of left ventricular (LV) systolic function for up to 30 weeks (Hoshijima et al., 2002). This demonstrated the potential use of rAAV as a vector for therapies involving progressive dilated cardiomyopathies and associated heart failure (Hoshijima et al., 2002).

Studies have demonstrated that AAV serotypes 1 and 2 are able to efficiently transduce the murine heart for up to 10 months following GFP transgene delivery (Du et al., 2004). Recently, we have found that intravenous delivery to the neonatal mouse of the newly characterized AAV serotype 9 results in β-galactosidase expression levels that are up to 50-fold higher in cardiac tissue than those obtained using AAV1 (Fig. 9.1) (Pacak, submitted 2004). The

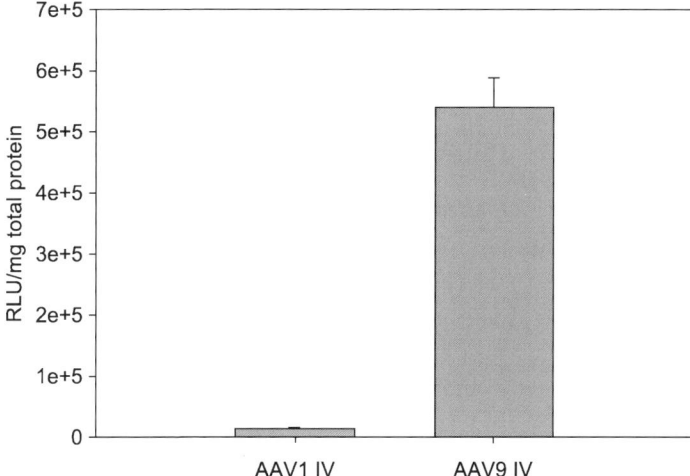

Fig. 9.1. β-Galactosidase enzyme activity in heart after IV injection of 1×10^{11} particles of rAAV-CMV-lacZ in mouse neonates 4 weeks post-injection. IV administration of AAV9 results in higher β-galactosidase expression in cardiac tissue than AAV1.

utilization of gene delivery vehicles that have a natural tropism for cardiac tissue would clearly be ideal for any genetic disease affecting solely that tissue type.

While the viral gene delivery systems do naturally possess many useful characteristics, there are limitations to each specific type of viral vector. The major consideration when using the retroviral and lentiviral delivery systems is their ability to randomly integrate into the host chromosome. Transgene size is a serious limitation for the AAV vectors, and the elicited immune response can be a problem when using Ad-mediated gene delivery. One issue for all recombinant vectors derived from viruses that are common infectious agents found within the general human population, is the potential for already-present neutralizing antibodies to hinder initial transduction efficiencies and prevent therapy effectiveness upon re-administration of the same viral serotype.

9.2. Non-viral gene delivery systems

9.2.1. DNA/DNA liposome complexes

Initial experiments utilizing non-viral-based gene delivery systems often involved simple injections of DNA directly into the myocardium (Lin et al., 1990; Acsadi et al., 1991). While expression was detected anywhere from 2 to 4 weeks post-injection, it would abruptly end due to what was believed to be an immune response to the high levels of naked DNA. However, when DNA was complexed to cationic liposomes, it was found to be far better tolerated (San et al., 1993) although transgene expression was variable. Other investigators have found that delivery of antisense c-myb oligonucleotides was able to successfully inhibit intimal arterial smooth muscle cell accumulation in a rat carotid injury model (Simons et al., 1992) demonstrating the potential for this therapeutic agent to be a useful short-term expression gene delivery system.

9.2.2. DNA/nanoparticles

More recently, investigators have demonstrated vascular gene delivery mediated by polyelectrolyte nanoparticles (Zaitsev et al., 2004). This method prevented particle aggregation under physiological conditions and resulted in high transfection efficiency into the rat carotid artery in vivo (Zaitsev et al., 2004). Non-viral delivery systems can successfully overcome many of the challenges faced by the viral systems such as viral toxicity, transgene size limitations, and potential for frequent random integration. Due to their typically short duration of expression, the non-viral systems are useful candidates for gene therapy applications requiring brief expression times. Although studies are being performed to extend this, presently the major limitations of the non-viral systems is the risk of potential binding to unintended target sequences and the fact that they are presently not capable of long-term persistent transgene expression.

9.3. Gene delivery route

One of the most important issues in cardiovascular gene therapy is the establishment of a clinically feasible gene delivery route. Transduction of the vasculature is often performed using a needle or catheter by simple intravenous (IV) or intraarterial (IA) injections of a gene delivery vehicle containing either a marker or therapeutic transgene of interest. The major challenge facing this delivery route is whether or not the gene delivery vehicle itself is able to transduce vascular and arterial walls. If a particular gene delivery system is unable to directly transduce the vasculature in vivo, it may be possible to use the system to transduce cells ex vivo and transplant them to the desired area of interest as previously mentioned with both retroviruses and lentiviruses.

Gene delivery to cardiac tissue is a significant challenge to the field of gene therapy. Direct intramyocardial injections using either a basic thoracotomy technique or catheter-based injection systems generally result in regional distribution of gene transfer (Fromes et al., 1999; Miao et al., 2000). Other techniques have been developed involving coronary injections or infusions of various vector-containing solutions in order to enhance diffusion of the gene delivery vehicle across the pericardial membrane (Greelish et al., 1999; Ikeda et al., 2002; Iwatate et al., 2003). Each of these methods requires an invasive surgery that adds to the accumulated risk of the potential human therapy. Ideally, a strategy involving a less-invasive delivery route combined with a serotype or tropism-determining moiety or nanoparticle capable of crossing vasculature would ultimately become a safer, more clinically relevant alternative to current approaches.

New technologies have also been developed to aid in the accuracy of the physical delivery of gene therapy vehicles. Magnetic resonance imaging has been utilized for in vivo real-time monitoring of catheter-based vascular gene delivery (Yang et al., 2001). This technique may prove particularly useful in situations where localized vascular gene delivery is desirable and more precise control over location of gene delivery is essential.

9.4. Cellular and gene therapy combinations

Another novel cardiovascular therapy that has recently come into play is the use of stem cell delivery. These approaches have thus far involved the use of gene delivery mechanisms to essentially force embryonic stem cells into cardiomyocyte differentiation with the goal of producing a cardiac pace-making cell (Arruda et al., 2004). Researchers are also developing methods for utilizing seeded adult mesenchymal stem cells as an implanted base to act as a depot for the delivery of localized gene therapies (Arruda et al., 2004).

One study compared direct delivery of an adenovirus vector encoding vascular endothelial growth factor (VEGF) to delivery of skeletal myoblasts, which had been transduced with Ad-VEGF. The results demonstrated that when each strategy was delivered 2 months post-left anterior descending coronary artery ligation in rats, the cellular delivery strategy increased neovascularization approximately 85% more than the Ad-mediated gene delivery alone (Askari et al., 2004). In addition, it was only with the cellular delivery strategy that an increase in cardiac function and shortening fraction were observed (Askari et al., 2004).

9.5. Conclusions

In general, there are a wide variety of circumstances in which some form of cardiovascular gene therapy can play an essential role in the management of the patient with cardiovascular disease. Some situations may call for the short-term expression of a gene in a specifically targeted region that may be best treated with a non-viral gene delivery system and a catheter-based delivery route. Other disease situations may require the long-term, global, persistent expression of a gene throughout the entire lifetime of the afflicted individual and could be best treated using a viral-based delivery system and a basic IV delivery route to achieve wide expression and biodistribution in the target tissue of myocardium (Table 9.1). All of these areas of research in cardiovascular gene delivery are carefully moving forward and patiently moving on the journey from bench to bedside in an attempt to one day establish cardiovascular gene delivery as an integral therapy in the clinical setting.

9.6. Methods

9.6.1. Direct intramyocardial injections in mouse neonates

In order to ensure accuracy of vector delivery to the heart and to prevent any potential injury to the animal from any movements during vector administration, neonates are anesthetized by induced hypothermia in which the animals are wrapped in cellophane and floated in an ice water bath until their activity slows. The neonate will be removed from ice and placed on a pad under a microscope and light, and rubbed down with a sterile alcohol pad. Up to 25 ul of vector can be directly injected into the heart through the chest cavity using a 33-gauge needle on a Hamilton syringe.

9.6.2. Direct intramyocardial injection in adult mice

Animals are anesthetized, intubated, and placed on a ventilator as previously described. The animal is sedated with a mixture of 2% isofluorane and O_2 (1–2 l) throughout the procedure. The animal while asleep, is placed on a pre-prepped surgical board and its limbs

TABLE 9.1
Summary table of various viral gene delivery systems and their characteristics

Gene delivery system	Transgene size capacity (Kizana and Alexander, 2003)	Duration of transgene expression	Persistence	Onset of transgene expression	Limitations
Retrovirus	7 kb	Long-term expression	Random integration into host chromosome	Early	Can only transduce dividing cells
Adenovirus	30 kb	Transient expression	Episomal (without rep gene)	Early (2 h) (Dumasius et al., 2003)	Restricted tropism, cytotoxic and immunogenic
Lentivirus	10 kb	Long-term expression	Random Integration into host chromosome	Early	Insertional mutagenesis concerns
Adeno-associated virus	4.8 kb	Long-term expression	Episomal	Late (>1 week for most serotypes)	Inefficient large-scale production techniques

are loosely restrained with tape, for the procedure. Hair on chest is clipped and fully removed using depelatory cream for 1 min, wiped with sterile saline then scrubbed three times with povidone–iodine scrub alternating with sterile saline. Upon reaching a surgical plane of anesthesia, bupivacaine is injected into the intercostal muscles to provide local anesthesia. Bupivacaine (0.25%) is diluted 1:10 in sterile water, saline, or PBS and injected at 0.1 ml/25 g mouse at site (http://iacuc.cwru.edu/policy/mouseaa.html). A 2 cm incision is made on the left side of the chest. A left thoracotomy is then performed and the area is kept open with retractors to expose the heart. An intramyocardial injection of 35 ul of virus is performed using a 33-gauge needle on a Hamilton syringe. Prior to complete closure of the chest wall, all air is evacuated from the thoracic cavity via suction. The chest wall is closed using interrupted 5-0 or 6-0 prolene suture and the skin is closed using interrupted 5-0 prolene suture. The mouse is then injected with .01–.02 ml/g of warm LRS or .9% sodium chloride IP or SC. The mouse is then monitored and respiration tube removed only after voluntary respiratory efforts are evident.

9.6.3. Adult cardiac-infusion delivery

Direct cardiac delivery into adults can be performed via the vector transfer procedure described by Ikeda et al. (2002). Briefly, animals are anesthetized with an intraperitoneal injection of sodium pentobarbital (75 mg/kg), and ventilated. The right carotid artery is then cannulated with a catheter placed at the aortic root. The animal is then placed under induced hypothermia until the body core temperature drops to below 26 °C. The ascending aorta and pulmonary arteries are occluded and histamine pre-treatment is delivered to the aorta (20 mmol/l, volume; 2.5 ul/g, body weight) for 3 min. The vector solution is then injected, the occlusions released 30 s later and the animal is resuscitated.

9.6.4. Pressure–volume relationships for cardiac contractility

These are terminal, non-recovery surgical procedures, which must be performed according to approved IACUC protocols. Subjects are anesthetized with either ketamine (100 mg/kg) and xylazine (5–10 mg/kg) or the anesthetic of choice for pressure–volume analyses in mice, a cocktail of urethan (1 mg/g), etomidate (0.02–0.025 mg/g), and morphine (0.001–0.002 mg/g). The animal is positioned supine on the operating table and body temperature is maintained throughout the procedure. The animal is then intubated and placed on a small animal ventilator at settings that provide normal ventilation for a 30 g animal. Recommended ventilator settings: breaths per minute: 105; tidal volume: .22 ml; inspiratory time: .33 s. (Settings provided in published papers are in the ranges of 70–20 breaths/min with tidal volumes of ~0.3–0.5 ml and may require adjustment for mice weighing less than 30 g.) The anesthesia machine is connected to the ventilator to keep the animal sedated with a mixture of 2% isofluorane and O_2 (1–2 l) throughout the remainder of the procedure. A small polyurethane catheter is placed in the left internal jugular vein via cut-down for infusion of pharmacologic agents. The heart is then exposed by a midline thoracotomy, and the chest walls retracted away from the midline. A 27G needle is used to make a stab wound in the apex of the heart. The pressure-conductance (PC) catheter is then inserted into the apex of the heart and passed anterograde through the aortic valve. Aortic flow is measured through placement of an ultrasound flow probe around the aortic root. Ventricular impedance, pressure, and aortic flow measurements are performed. Data can be collected using custom graphing software from which pressure–volume relationships are plotted. Pharmacological alteration in contractility can be produced through administration of dobutamine. Reduction of preload can be accomplished by transient occlusion of the inferior vena cava. From these results a set of pressure–volume loops are generated allowing for determination of the end-systolic pressure–volume relationship (ESPVR) (Georgakopoulos and Kass, 2004).

References

Acsadi, G., Jiao, S. S., Jani, A., Duke, D., Williams, P., Chong, W. and Wolff, J. A. (1991). Direct gene transfer and expression into rat heart in vivo. New Biol. *3*, 71–81.

Arruda, V. R., Schuettrumpf, J., Herzog, R. W. et al. (2004). Safety and efficacy of factor IX gene transfer to skeletal muscle in murine and canine hemophilia B models by adeno-associated viral vector serotype 1. Blood *103(1)*, 85–92.

Asfour, B., Baba, H. A., Scheld, H. H. et al. (2002). Uniform long-term gene expression using adeno-associated virus (AAV) by ex vivo recirculation in rat-cardiac isografts. Thorac Cardiovasc. Surg. *50(6)*, 347–350.

Askari, A., Unzek, S., Goldman, C. K., Ellis, S. G., Thomas, J. D., Di Corleto, P. E., Topol, E. J. and Penn, M. S. (2004). Cellular, but not direct, adenoviral delivery of vascular endothelial growth factor results in improved left ventricular function and neovascularization in dilated ischemic cardiomyopathy. J. Am. Coll. Cardiol. *19*, 1908–1914.

Byun, J. H., Huh, J. E., Park, S. J. et al. (2000). Myocardial injury-induced fibroblast proliferation facilitates retroviral-mediated gene transfer to the rat heart in vivo. J. Gene Med. *2(1)*, 2–10.

Chao, H. J., Liu, Y. B., Rabinowitz, J. et al. (2000). Several log increase in therapeutic transgene delivery by distinct adeno-associated viral serotype vectors. Mol. Ther. *2(6)*, 619–623.

Du, L. L., Kido, M., Lee, D. V. et al. (2004). Differential myocardial gene delivery by recombinant serotype-specific adeno-associated viral vectors. Mol. Ther. *10(3)*, 604–608.

Dumasius, V., Jameel, M., Burhop, J. et al. (2003). In vivo timing of onset of transgene expression following adenoviral-mediated gene transfer. Virology *308(2)*, 243–249.

Frimpong, K. and Spector, S. A. (2000). Cotransduction of nondividing cells using lentiviral vectors. Gene Ther. *7(18)*, 1562–1569.

Fromes, Y., Salmon, A., Wang, X. et al. (1999). Gene delivery to the myocardium by intrapericardial injection. Gene Ther. *6(4)*, 683–688.

Georgakopoulos, D. and Kass, D. A. (2004). Minimal force-frequency modulation of inotropy and relaxation of in situ murine heart. J. Physiol. *15*, 535–545.

Gnatenko, D., Arnold, T. E., Zolotukhin, S. et al. (1997). Characterization of recombinant adeno-associated virus-2 as a vehicle for gene delivery and expression into vascular cells. J. Investig. Med. *45(2)*, 87–98.

Greelish, J. P., Su, L. T., Lankford, E. B. et al. (1999). Stable restoration of the sarcoglycan complex in dystrophic muscle perfused with histamine and a recombinant adeno-associated viral vector. Nat. Med. *5(4)*, 439–443.

Guzman, R. J., Lemarchand, P., Crystal, R. G. et al. (1993). Efficient gene-transfer into myocardium by direct-injection of adenovirus vectors. Circul. Res. *73(6)*, 1202–1207.

Hauck, B., Chen, L. and Xiao, W. D. (2003). Generation and characterization of chimeric recombinant AAV vectors. Mol. Ther. *7(3)*, 419–425.

Haviernik, P. and Bunting, K. D. (2004). Safety concerns related to hematopoietic stem cell gene transfer using retroviral vectors. Curr. Gene Ther. *4(3)*, 263–276.

Hoshijima, M., Ikeda, Y., Iwanaga, Y. et al. (2002). Chronic suppression of heart-failure progression by a pseudophosphorylated mutant of phospholamban via in vivo cardiac rAAV gene delivery. Nat. Med. *8(8)*, 864–871.

Ikeda, Y., Gu, Y., Iwanaga, Y. et al. (2002). Restoration of deficient membrane proteins in the cardiomyopathic hamster by in vivo cardiac gene transfer. Circulation *105(4)*, 502–508.

Iwatate, M., Gu, Y., Dieterle, T. et al. (2003). In vivo high-efficiency transcoronary gene delivery and Cre-LoxP gene switching in the adult mouse heart. Gene Ther. *10(21)*, 1814–1820.

Kessler, P. D., Podsakoff, G. M., Chen, X. et al. (1996). Gene delivery to skeletal muscle results in sustained expression and systemic delivery of a therapeutic protein. Proc. Natl. Acad. Sci. USA *93*, 14082–14087.

Kizana, E. and Alexander, I. E. (2003). Cardiac gene therapy: Therapeutic potential and current progress. Curr. Gene Ther. *5*, 418–451.

Lemarchand, P., Jones, M., Yamada, I. and Crystal, R. G. (1993). In vivo gene-transfer and expression in normal, uninjured blood-vessels using replication deficient recombinant adenovirus vectors. Clin. Res. *41(2)*, A202.

Li, J., Wang, D., Qian, S. et al. (2003). Efficient and long-term intracardiac gene transfer in delta-sarcoglycan-deficiency hamster by adeno-associated virus-2 vectors. Gene Ther. *10(21)*, 1807–1813.

Lin, H., Parmacek, M. S., Morle, G., Bolling, S. and Leiden, J. M. (1990). Expression of recombinant genes in myocardium in vivo after direct injection of DNA. Circulation *82*, 2217–2221.

Lynch, C. M., Clowes, M. M., Osborne, W. R. A. et al. (1992). Long-term expression of human adenosine-deaminase in vascular smooth-muscle cells of rats—A model for gene-therapy. Proc. Natl. Acad. Sci. USA *89(3)*, 1138–1142.

Miao, W. F., Luo, Z. Y., Kitsis, R. N. and Walsh, K. (2000). Intracoronary, adenovirus-mediated Akt gene transfer in heart limits infarct size following ischemia-reperfusion injury in vivo. J. Mol. Cell. Cardiol. *32(12)*, 2397–2402.

Newman, K. D., Nguyen, N. and Dichek, D. A. (1991). Quantification of vascular graft seeding by use of computer-assisted image-analysis and genetically modified endothelial-cells. J. Vasc. Surg. *14(2)*, 140–146.

Nicklin, S. A., White, S. J., Buning, H. et al. (2003). Site-specific gene delivery to the vasculature in vivo using peptide-targeted adeno-associated virus (AAV). vectors. Atherosclerosis *169(1)*, 200.

O'Donnell, J. M., Sumbilla, C. M., Ma, H. et al. (2001). Tight control of exogenous SERCA expression is required to obtain acceleration of calcium transients with minimal cytotoxic effects in cardiac myocytes. Circ. Res. *88(4)*, 415–421.

Psarras, S., Karagianni, N., Kellendonk, C. et al. (2004). Gene transfer and genetic modification of embryonic stem cells by Cre- and Cre-PR-expressing MESV-based retroviral vectors. J. Gene Med. *6(1)*, 32–42.

Roks, A. J. M., Pinto, Y. M., Paul, M. et al. (1997). Vectors based on Semliki Forest virus for rapid and efficient gene transfer into non-endothelial cardiovascular cells: Comparison to adenovirus. Cardiovasc. Res. *35(3)*, 498–504.

San, H., Yang, Z. Y., Pompili, V. J., Jaffe, M. L., Plautz, G. E., Xu, L., Felgner, J. H., Wheeler, C. J., Felgner, P. L., Gao, X. et al. (1993). Safety and short-term toxicity of a novel cationic lipid formulation for human gene therapy. Hum. Gene Ther. *4*, 781–788.

Simons, M., Edelman, E. R., De Keyser, J. L., Langer, R. and Rosenberg, R. D. (1992). Antisense c-myb oligonucleotides inhibit intimal arterial smooth muscle cell accumulation in vivo. Nature *359*, 67–70.

Stratfordperricaudet, L. D., Makeh, I., Perricaudet, M. and Briand, P. (1992). Widespread long-term gene-transfer to mouse skeletal-muscles and heart. J. Clin. Inves. *90(2)*, 626–630.

Verhoeyen, E. and Cosset, F. L. (2004). Surface-engineering of lentiviral vectors. J. Gene Med. *6*, S83–S94.

Waddington, S. N., Mitrophanous, K. A., Ellard, F. M. et al. (2003). Long-term transgene expression by administration of a lentivirus-based vector to the fetal circulation of immuno-competent mice. Gene Ther. *10(15)*, 1234–1240.

Warrington, K. H., Gorbatyuk, O. S., Harrison, J. K. et al. (2004). Adeno-associated virus type 2 VP2 capsid protein is nonessential and can tolerate large peptide insertions at its N terminus. J. Virol. *78(12)*, 6595–6609.

Yang, X., Atalar, E., Li, D., Serfaty, J., Wang, D., Kumar, A. and Cheng, L. (2001). Magnetic resonance imaging permits in vivo monitoring of catheter-based vascular gene delivery. Circulation *104*, 1588–1590.

Yao, S. N., Wilson, J. M., Nabel, E. G. et al. (1991). Expression of human factor-Ix in rat capillary endothelial-cells—Toward somatic gene-therapy for hemophilia-B. Proc. Natl. Acad. Sci. USA *88(18)*, 8101–8105.

Yee, J. K. and Zaia, J. A. (2001). Prospects for gene therapy using HIV-based vectors. Somat. Cell. Mol. Genet. *26*, 159–174.

Yue, Y. P., Li, Z. B., Harper, S. Q. et al. (2003). Microdystrophin gene therapy of cardiomyopathy restores dystrophin–glycoprotein complex and improves sarcolemma integrity in the mdx mouse heart. Circulation *108(13)*, 1626–1632.

Zaitsev, S., Cartier, R., Vyborov, O., Sukhorukov, G., Paulke, B. R., Haberland, A., Parfyonova, Y., Tkachuk, V. and Bottger, M. (2004). Polyelectrolyte nanoparticles mediate vascular gene delivery. Pharm Res. *21(9)*, 1656–1661.

Zhao, J., Pettigrew, G. J., Thomas, J. et al. (2002). Lentiviral vectors for delivery of genes into neonatal and adult ventricular cardiac myocytes in vitro and in vivo. Basic Res. Cardiol. *97(5)*, 348–358.

CHAPTER 10

Gene therapy for lysosomal storage disorders

Kerry O. Cresawn and Barry J. Byrne

Department of Pediatrics, Powell Gene Therapy Center, University of Florida, Box 100296, Gainesville, FL 32610-0296, USA

10.1. The lysosome

Lysosomes are acidic organelles contained within nearly all eukaryotic cells. The over 40 hydrolytic enzymes contained within the lysosome play a crucial role in maintaining cell homeostasis by breaking down a variety of materials, many of which yield by-products that can be reutilized for metabolism and cell renewal. Microorganisms, intracellular macromolecules (including organelles, lipids, proteins, and carbohydrates), and appropriately tagged extracellular macromolecules are all potential targets for lysosomal degradation. Genetic mutations resulting in lysosomal enzyme deficiency lead to progressive accumulation of these undegraded substrates within the lysosomes and consequential engorgement of the organelles. This leads to compromised architecture and damage at the cellular and tissue level leading to organ dysfunction, and in many cases, early death.

10.2. Lysosomal storage diseases

There are over 40 lysosomal storage disorders (LSDs) characterized by the specific enzyme deficiency and accumulated substrate. Pathologies associated with LSDs are multisystemic and variable including CNS, skeletal, cardiovascular, renal, and ocular system involvement. The aggregate incidence is estimated to approach 1 in 7000 live births (Ellinwood et al., 2004). Inheritance for LSDs is primarily autosomal recessive with the exception of two X-linked diseases (Fabry and mucopolysaccharidosis (MPS) II). Treatment for LSDs relies on providing functional enzyme to the lysosomes of affected cells and has traditionally been confined to bone marrow transplantation, and enzyme replacement therapy (ERT).

10.3. Current therapies

The most commonly used therapy to treat LSDs is heterologous bone marrow transplantation (BMT). This treatment provides both normal bone marrow and bone marrow-derived cells, which release enzyme continuously. Unfortunately, BMT is associated with several problems and risks including the availability of a suitable donor, poor response to therapy, and sustained immune suppression. BMT therapies for MPS I, MPS II, MPS III, metachromatic leukodystrophy, and non-neuronopathic forms of Gaucher disease have demonstrated promising results. In most successful cases, the pathology is reversed in the visceral organs with variable or unclear success in the CNS (Laine et al., 2004).

Success with enzyme replacement therapy has been demonstrated in patients for the treatment of non-neuronopathic forms of Gaucher disease (Weinreb et al., 2002) and Fabry disease (Eng et al., 2001b). Clinical trials are currently underway evaluating the use of ERT to treat a variety of LSDs including glycogen storage disease type II (GSDII) (Pompe disease), MPS I (Kakkis et al., 2001; Wraith et al., 2004), MPS II (Muenzer et al., 2002), MPS VI

(Harmatz et al., 2004), and Fabry disease (Eng et al., 2001a,b). ERT success is generally restricted to LSDs without CNS involvement, as systemically administered enzyme is unable to cross the blood–brain barrier (BBB) and neurological correction would require routine direct injections into the brain. Furthermore, because of the short half-life of the enzyme, regular bolus dosing is required. This increases the potential for an immune response to the protein, especially in null patients. Additionally, the expense associated with routine protein administration and the high cost to manufacture sufficient quantities of purified protein can create a financial burden for the patients, their families, and healthcare providers. Other therapies previously used or under investigation for treatment of LSDs include diet modifications (Bodamer et al., 1997, 2000), substrate reduction therapy (Ellinwood et al., 2004), and pharmacological chaperones (Asano et al., 2000; Sawkar et al., 2002; Fan, 2003).

10.4. Gene therapy

Gene therapy is an attractive alternative to ERT for several reasons. Gene therapy offers the potential for a sustained therapeutic effect from a single administration of vector. In contrast to ERT, this single treatment decreases (although does not eliminate) the probability of an immune response to the enzyme and would also decrease the cost to the patient. LSDs are ideal candidates for gene therapy treatment for many reasons. LSDs are generally well-characterized, single gene disorders. Many genetically engineered mouse models as well as naturally occurring animal models are available for understanding the disease pathophysiology and evaluating gene therapy treatments. While the amount of enzyme required to correct the disease varies depending on the enzyme deficiency and affected tissues, restoring enzyme levels to those observed in unaffected heterozygote patients (50% of normal) or patients with a milder LSD phenotype (10–20% of normal) should significantly improve survival and quality of life (Reuser et al.,

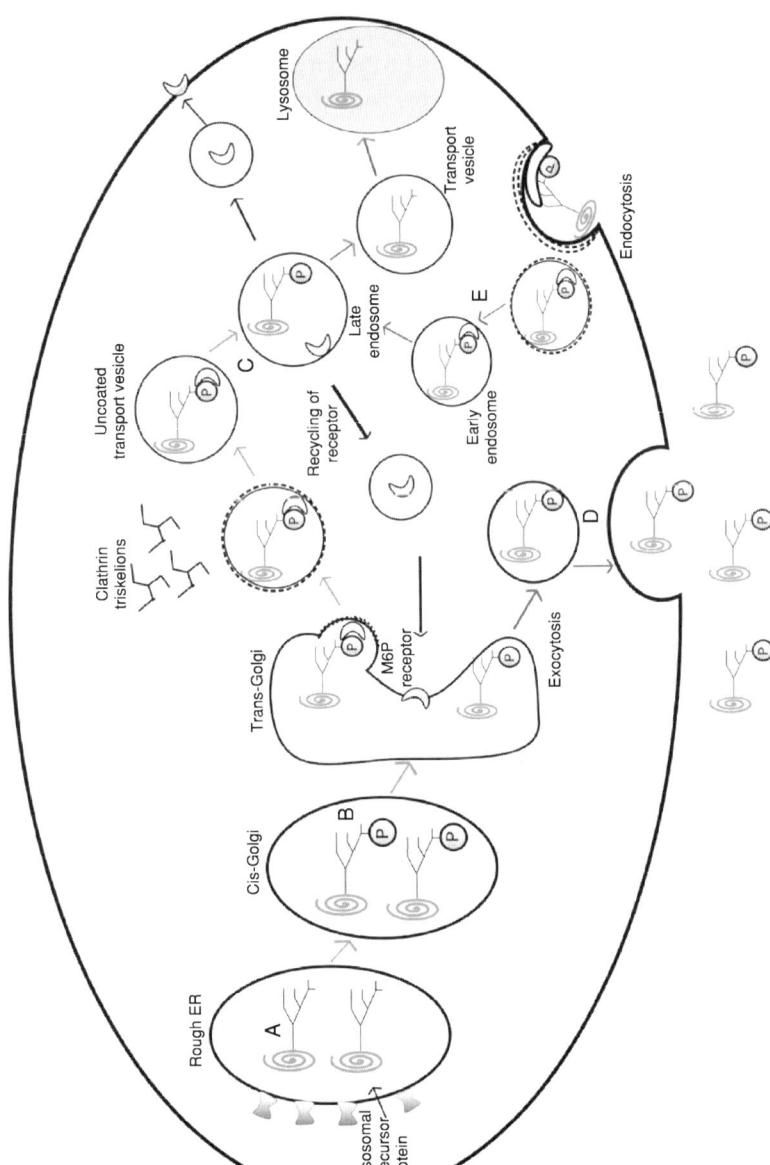

Fig. 10.1. Lysosomal enzymes are synthesized as precursor proteins that undergo extensive modifications during transport to the lysosome. Newly synthesized lysosomal proteins have an amino-terminal signal sequence that mediates

1995). Therefore, complete restoration of enzyme levels will not be necessary. Tight regulation of lysosomal enzyme production will also not be necessary due to the low pH requirement for enzyme activity (rendering cytoplasmic or circulating enzymes inactive at neutral pH). Most importantly, a majority of newly synthesized lysosomal enzymes traffic to the lysosomes. However, a small percentage of the enzyme is secreted from the cell, and can be taken up by other cells via receptor-mediated endocytosis and trafficked to the lysosomes of those cells (Reuser et al., 1995) (Fig. 10.1). This ability of cells to both secrete newly synthesized enzyme and endocytose circulating enzyme allows for gene therapy-mediated correction of multiple tissues either by systemic vector administration or direct vector administration to one tissue (i.e., liver or muscle), which will in turn synthesize and secrete functional enzyme for another.

---

cotranslational transport into the lumen of the endoplasmic reticulum (ER). After cleavage of the signal peptide, the nascent enzyme is N-glycosylated by the addition of oligosaccharide groups to asparagines residues (A). Upon transport to the *cis*-Golgi, the carbohydrate chains are modified resulting in phosphorylated mannose residues (B). The resulting mannose 6-phosphate groups play a critical role in lysosomal sorting. The newly synthesized M6P-containing proteins recognize M6P receptors present on the Golgi. The protein-receptor complex is transported from the Golgi to the lysosome via clathrin coated vesicles that bud off from the *trans*-Gogi and fuse to endosomes. The lower pH of the late endosomal compartments allows the receptor–protein complex to dissociate (C). The protein is trafficked to the lysosome while the receptor is recycled either back to the Golgi or to the plasma membrane. This pathway occurs for approximately 80–90% of synthesized lysosomal proteins. The remaining 10–20% can alternatively be routed from the Golgi to the plasma membrane via secretory granules, enter the extracellular fluid (D), bind M6P receptors on the cell surface of other cells, internalize, and trafficked through endosomal compartments to the lysosomes (E).

10.5. Glycogen storage disease type II

Among the 40 LSDs characterized, gene therapy-mediated correction has been reported in animal models for at least 17 (Table 10.1). Gene therapy progress for many more LSDs is currently in the early stages of in vitro studies with promising results for animal studies in the near future. Critical differences among the LSDs, such as the presence or absence of CNS involvement, will greatly impact the gene therapy-mediated treatment approach and outcome. However, the characteristic biogenesis and trafficking feature of all lysosomal enzymes does allow for lessons from gene therapy studies of one disease to be broadly applied to others. With the exception of discussion on gene therapy treatment for CNS pathologies, this chapter will focus on gene therapy studies performed for the treatment of glycogen storage disease type II. Lessons from these studies will illustrate the issues currently important for the treatment of LSDs including choice of vector, target organ, delivery method, and age of treatment as well as important immunology concerns.

GSDII (Pompe disease, acid-maltase deficiency) is caused by a complete or partial deficiency of the enzyme responsible for hydrolysis of lysosomal glycogen, acid α-glucosidase (GAA). GAA deficiency leads to glycogen accumulation in all tissues with severe pathologies in the cardiac and skeletal muscles. The severity of GSDII is ultimately related to the degree of enzyme deficiency and represents a continuum of clinical disease. The most severe variant is the early-onset form, also known as Pompe disease. Pompe patients have complete or near-complete GAA deficiency and usually die within the first two years of life due to cardiorespiratory failure. Later-onset GSDII occurs anywhere from early childhood to late adulthood and is primarily characterized by skeletal muscle weakness, respiratory insufficiency, and absent or mild cardiac involvement (Hirschhorn and Reuser, 2000; Raben et al., 2002). To date, there has been no reported CNS involvement in GSDII. However, correction of cardiac and skeletal

TABLE 10.1
Gene therapy progress for lysosomal storage disorders as of 2004

Disease	Enzyme deficiency	Substrate accumulated	Major manifestations	Animal models	In vivo vectors and routes tested
Fabry	α-Galactosidase a (α-gal a)	α-Galactosyl-terminal lipids	Renal failure, pain, skin lesions, strokes, cardiomyopathy	Mouse	Retrovirus: Ex vivo (Takenaka et al., 2000) AAV: PV (Jung et al., 2001); IV (Park et al., 2003; Ziegler et al., 2004); IM (Takahashi et al., 2002) Ad: Inhalation (Li et al., 2002); IV (Ziegler et al., 1999)
Gaucher	Acid β-glucosidase	Glycosphingo-lipids	Hepatosplenomegaly, skeletal disease, pancytopenia, neurological (type II)	Mouse	Ad: IV, IN (Marshall et al., 2002) Plasmid: IV, IN (Marshall et al., 2002) Lentivirus: PV, IV (Kim et al., 2004) Retrovirus: Ex vivo (Dunbar et al., 1998; Liu et al., 1998)
GSD II (Pompe)	Acid α-glucosidase	Glycogen	Cardiomegaly, muscle weakness, respiratory insufficiency	Mouse, quail, cattle, cat, sheep	AAV: IV (Hong et al., 2004) Ad: IV (Ding et al., 2001, 2002); IM (Pauly et al., 1998; Ding et al., 2002; Martin-Touaux et al., 2002); Intra-cardiac (Pauly et al., 1998); PV (Pauly et al., 2001)

(*continued*)

TABLE 10.1 (*continued*)

Disease	Enzyme deficiency	Substrate accumulated	Major manifestations	Animal models	In vivo vectors and routes tested
MPS I (Hurler)	α-L-Iduronidase	Dermatan sulfate, heparan sulfate	Skeletal, organomegaly, cardiovascular, neurological, ocular	Cat, dog, mouse	AAV: IM, Intracardiac (Fraites et al., 2002); IV (Sun et al., 2003); PV (Cresawn et al., in press; Sun et al., 2003); In utero (Rucker et al., 2004) Plasmid: Particle bombardment (Martiniuk et al., 2002)
MPS II (Hunter)	Iduronate sulfatase	Dermatan sulfate, heparan sulfate	Skeletal, bone, organomegaly, neurological	Mouse, dog	Retrovirus: Ex-vivo (Zheng et al., 2004); in utero (Meertens et al., 2002) Plasmid: IM (Lutzko et al., 1999) AAV: Intracranial (Desmaris et al., 2004); IV (Hartung et al., 2004) Plasmid: IM (Tomanin et al., 2002)
MPS IIIB (Sanfilippo)	α-N-Acetyl-glucosaminidase	Heparan sulfate	Neurological	Mouse	AAV: Intracranial (Fu et al., 2002)

Disease	Enzyme	Substrate	Symptoms	Animal model	Gene therapy approach
MPS VI (Maroteaux-Lamy)	N-Acetyl galactosamine 4-sulfatase	Dermatan sulfate	Skeletal, bone, ocular	Mouse, rat, cat, dog	Retrovirus: Ex vivo (Simonaro et al., 1999)
MPS VII (Sly)	β-Glucoronidase	Dermatan sulfate, heparan sulfate, chondroitin 4-, 6-sulfates	Hepatosplenomegaly, skeletal, neurological, cardiovascular, ocular	Mouse, dog, cat	Retroviral: Ex-vivo, IV Adenoviral: IV, intracorneal, intraventricular, intrastriatal AAV: IV, IM, intracorneal, intrathecal, intraventricular, intracranial Lentivirus: IV, intracerebral, intra-striatal All reviewed in Ellinwood et al. (2004). Ad: IV (Du et al., 2002)
Wolman	Acid lipase	Cholesterol esters	Hepatosplenomegaly, steatorrhea, GI	Mouse	
Niemann-Pick A (neuropathic) and B and B	Acid-sphingomyelinase	Sphingomyelin	Hepatosplenomegaly, a: neurological, b: respiratory, cardiomegaly,	Mouse	Retroviral: Ex-vivo (Miranda et al., 2000; Jin and Schuchman, 2003)
Krabbe	Galactosylceramidase	Galactosylsphingosine	Neurological	Mouse dog, cat sheep	Ad: Intraventricular (Shen et al., 2001)
Metachromatic leukodystrophy	Arylsulfatase a	Sulfated glycolipids	Neurological, hyoptonia, ocular	Mouse	Retrovirus: Ex vivo (Matzner et al., 2000a,b, 2002) Lenivirus: Intracranial (Consiglio et al., 2001)

(continued)

TABLE 10.1 (continued)

Disease	Enzyme deficiency	Substrate accumulated	Major manifestations	Animal models	In vivo vectors and routes tested
Ceroid lipofuscinosis	Palmitoyl protein thioesterase-1 (CNL1), tripeptidyl peptidase (CNL2), membrane proteins with unknown function (CNL3, 5, 8)	Lipopigments	Neurological	Mouse, dog, sheep	AAV: Intracranial (Haskell et al., 2003; Griffey et al., 2004) Ad: Intracranial (Haskell et al., 2003) Lentivirus: Intracranial (Haskell et al., 2003)
GM_1 gangliosidosis	Acid β-galactosidase	gm1 gangliosides	Neurological	Mouse	Ad: IV (Takaura et al., 2003)
Tay Sachs	β-Hexosaminidase	gm2 gangliosides	Neurological	Mouse	Ad: IV (Guidotti et al., 1999) Retroviral: Ex vivo (Lacorazza et al., 1996)
Sandhoff disease	Hexominidase a and b	gm2 gangliosides	Neurological	Mouse	Plasmid: IV (Yamaguchi et al., 2003) Ad: Intracerebral (Bourgoin et al., 2003)
Galactosialidosis	N-Acetyl-α-neuraminidase, β-galactosidase	Sialyloligosaccharides	Neurological, hepatosplenomegaly, renal, skeletal, opthamologic, dysotosis multiplex, cardiovascular	Mouse	Retrovirus: Ex vivo (Leimig et al., 2002)

IV: intravenous, PV: poral vein, IN: intranasal, IM: intramuscular.

muscles of the more severely affected Pompe patients (resulting in survival to adulthood) may reveal unforeseen neurological consequences.

GAA is synthesized as a 110 kDa precursor protein, which is core-glycosylated in the ER and acquires mannose 6-phosphate (M6P) residues in the Golgi as diagrammed in Fig. 10.1. In the lysosomal compartments, the precursor protein is processed to yield an intermediate 95 kDa form and the two mature, fully active, 76 and 70 kDa, forms of GAA (Reuser et al., 1995; Raben et al., 2002). The 110 kDa form is secreted and can be detected in the media of infected cells or serum of transduced animals and the remaining forms can be found in expressing cells.

Several naturally occurring GSDII animal models have been reported (Walvoort, 1983). However the Japanese quail is the only model that has been successfully maintained and has even been used for gene therapy studies (Lin et al., 2002). Three different genetically engineered GSDII mouse models have been created by targeted disruption of the GAA gene. The two most extensively used models for studying gene replacement therapies are the homozygous exon 13 knockout (Bijvoet et al., 1998) and homozygous exon 6 knockout models (Raben et al., 1998), both created by insertion of a neomycin gene cassette in the respective exon. The studies discussed in this chapter predominately use the exon 6 knockout model which has been biochemically, histologically, and functionally characterized (Raben et al., 1998, 2000; Fraites et al., 2002; Rucker et al., 2004). The exon 6 knockout model (Gaa–/–) has no residual GAA activity and glycogen accumulation is evident as early as 1 month of age (Rucker et al., 2004). These mice also recapitulate the functional phenotype of the patient population with skeletal muscle weakness (Fraites et al., 2002) and respiratory insufficiency (Rucker et al., 2004). Functional assessment of the cardiac function in these mice is still under investigation. However, preliminary echo-cardiography results suggest an age-related increase in left ventricular mass index beginning at 6 months of age (unpublished results).

10.6. Gene therapy for GSD II: Proof of concept studies

The first study to demonstrate correction of GAA-deficient cells from a viral vector, involved an in vitro transduction of myoblasts and fibroblasts isolated from a patient with Pompe disease (Zaretsky et al., 1997). Using a retroviral vector carrying the GAA cDNA, there was an increase in GAA activity and a decrease in glycogen content in both of these cell types. More importantly, the transduced cells were able to cross-correct GAA-deficient myoblasts by secretion of GAA. Since this original study, there have been no reported studies using retroviral-mediated gene delivery for the treatment of GSDII. Majority of the work has been performed with adenoviral and adeno-associated viral (AAV) vectors and a few studies with plasmid-mediated gene delivery (Martiniuk et al., 2002).

The first adenoviral (Ad)-mediated gene delivery studies for GSDII were performed in GSDII patient fibroblasts, myoblasts, and myotubes (Nicolino et al., 1998; Pauly et al., 1998, 2001). Several important findings came from these early studies including the ability to achieve overexpression of GAA from Ad vectors of up to 19–20-fold over untreated normal cells (Nicolino et al., 1998; Pauly et al., 1998), localization of the Ad-delivered GAA protein to the lysosomes (Pauly et al., 1998, 2001), clearance of accumulated glycogen in treated cells (Nicolino et al., 1998; Pauly et al., 1998), secretion of the 110 kDA precursor form into the culture media (Nicolino et al., 1998; Pauly et al., 1998), and M6P receptor-mediated uptake of GAA secreted from transduced cells by GAA-deficient cells (Nicolino et al., 1998; Pauly et al., 1998).

With proof of concept established that viral vector-delivered GAA can restore biochemical and histological phenotype of affected cells and cross-correct untransduced cells, these studies were quickly followed by in vivo assessment of Ad-mediated GAA delivery in animals. Intracardiac and intramuscular delivery of Ad-human GAA (hGAA) was performed in newborn rats by Pauly et al. (1998) resulting in 10- and 6-fold normal levels of GAA in injected tissues, respectively. The first demonstration of cross-correction of untransduced, affected muscle tissues from liver-produced hGAA was reported by Amalfitano et al. (1999) and Pauly et al. (2001). The IV delivery studies performed by Amalfitano et al. involved administering a modified Ad vector expressing hGAA under control of the cytomegalovirus (CMV) promoter to the reto-orbital sinus at a dose of 1×10^9 PFU. This resulted in liver-restricted transgene expression with superphysiologic levels in the plasma, cardiac, and skeletal muscles. A decrease in plasma GAA activity occurred from 3 to 12 days post-injection and was suggestive of an immune response to the vector and/or protein and/or downregulation of the CMV promoter in hepatic tissue Amalfitano et al. (1999). Adenoviral-mediated, intravenous delivery of hGAA performed by Pauly et al. (2001) involved portal vein delivery at the same dose and also resulted in liver-restricted GAA gene expression with superphysiologic levels in the heart and skeletal muscle. These original cross-correction studies and others (Ding et al., 2001, 2002; Martin-Touaux et al., 2002) for correction of GSDII reveal the importance of the alternative secretory pathway of lysosomal enzyme for gene therapy-mediated correction. Despite the small percentage of newly synthesized lysosomal enzymes that is secreted from the cell (10–20%), 100-fold overexpression in the liver achieved from a single intravenous injection of Ad-hGAA vectors was able to restore the biochemical phenotype in the affected muscle tissues.

Adenoviral vectors are currently being used in clinical trials to treat a variety of diseases including cancer and inborn errors of metabolism. Genetic modifications of adenoviral vectors involving deletion of one, many, or all genes (Robbins et al., 1998) have improved the safety of adenoviral vectors. However, toxicity concerns of adenoviral vectors which were highlighted by the death of a gene therapy clinical trial patient in 1999 (Raper et al., 2003) have encouraged the use of an alternative viral vector that is not associated with any human disease, adeno-associated virus.

10.7. Recombinant adeno-associated virus vector studies

The safety profile of recombinant AAV (rAAV) vectors is attributed to the lack of association with human disease as well as the ability to generate recombinant vectors that contain no viral genes. Similar to adenoviral vectors, AAV can infect both dividing and non-dividing cells, and has a broad tissue tropism. In contrast to adenoviral vectors, rAAV vectors have a limited packaging capacity of about 4.9 kb, limiting their use to diseases in which the deficient gene is relatively small.

Most of the understanding about the basic biology and vector applications of AAV has been achieved studying AAV serotype 2 vectors. Currently, at least nine distinct AAV serotypes have been isolated and characterized (Gao et al., 2002). The standard approach to generating rAAV vectors of alternative serotypes is pseudotyping, which involves packaging the gene expression cassette with AAV2 inverted terminal repeat (ITR) elements in the capsid of the alternative serotypes (Zolotukhin et al., 2002). In vitro and in vivo studies comparing the different serotypes have revealed important information about optimal serotypes for targeting specific tissues. The results from these studies are important for understanding the appropriate approach to directly treat the affected tissues of the many LSDs as they vary from muscle to CNS to renal. However, the possibility to treat the major manifestations of many LSDs by a cross-correction mechanism (due to the characteristic secretion/receptor-mediated uptake pathway) suggests that serotypes that are most efficient in transduction of the two depot organs (liver and skeletal muscle) would be suitable for most LSD gene therapy models. AAV serotypes 1 and 6 have demonstrated the highest levels of expression in the muscle, whereas serotypes 5, 7, and 8 have proven to be most efficient in the liver (Xiao et al., 1999; Gao et al., 2002; Mingozzi et al., 2002). An important exception to this cross-correction model are the many LSDs with neurological involvement, including but not limited to Gaucher type II, Tay-Sachs, ceroid lipofuscinosis, Niemann-Pick A disease, MPS I,

II, and IIIB, in which direct cranial injections have proven to be the most successful approach to treat the CNS pathologies in animal models.

10.8. Recombinant AAV-mediated treatment of GSDII

Several different methods of rAAV-mediated correction of GSDII have been reported. The first group of studies established both that rAAV-delivered GAA can overexpress GAA in transduced cells in vitro with cross-correction of GAA-deficient cells and that direct intramuscular or intramyocardial injections of rAAV-GAA can restore GAA levels to near normal in injected tissues (Fraites et al., 2002). After intramuscular injection of rAAV2 and rAAV1-mGAA vectors, Fraites et al. observed 18-fold greater levels of expression with rAAV1 vectors than with rAAV2 vectors in the injected tissue, with rAAV1 vectors yielding 450-fold over normal levels of GAA. The restored enzyme levels in the injected skeletal muscle resulted in glycogen clearance and muscle function improvement. Intramyocardial delivery with rAAV2 vectors resulted in 70% of normal GAA levels in the heart (Fraites et al., 2002).

The cardiac and hind-limb muscles are two of the three primary targets for correction of GSDII. However, respiratory failure is the lead cause of mortality in later-onset patients and a contributing factor of mortality in the early-onset patients. Therefore, correction of the diaphragm muscle is critically important as well. Direct transduction of the diaphragm by conventional injection is not feasible in mice as the diaphragm is approximately six cells thick. As an alternative approach, Mah et al. (2004) demonstrated direct diaphragm transduction using a novel gel-based method for vector delivery. Recombinant AAV vector diluted in saline was added to a water soluble, glycerine-based gel, mixed by vortexing and applied to the diaphragm of anesthetized Gaa–/– mice. The vector-gel solution was allowed to sit on the diaphragm for 5 min before closing the animal. Six weeks post-delivery, the authors observed

120% of normal GAA activity and significant clearance of glycogen in the diaphragm of vector-gel treated mice compared to only 50% of normal GAA activity and no detectable glycogen clearance in mice treated with virus alone.

A different approach to diaphragm transduction was performed by Rucker et al. (2004). This method involved in utero delivery of rAAV-hGAA vectors to fetal Gaa–/– mice at 15 days post-coitus. Gaa–/– mice were analyzed at 1 month of age and results revealed 100% of normal activity in the diaphragm, complete clearance of glycogen, restoration of muscle fiber architecture, and improvement in contractile function. This method of delivery not only corrected one of the critical pathologies of GSDII but was the first example of early intervention of GSDII. The functional phenotype of Gaa–/– mice deteriorates with age, as glycogen accumulation increases and permanent damage of muscle fibers occurs. At a certain age it is likely that the pathologies will be irreversible. Therefore, early intervention, perhaps in utero or in neonatal life, will be important.

Due to the systemic muscle pathology of GSDII, transduction of multiple tissues or transduction of a depot organ with cross-correction to other tissues will be necessary to correct the phenotype. To evaluate rAAV-mediated cross-correction of GSDII previously demonstrated with adenoviral vectors, Cresawn et al. (in press) delivered rAAV5 and eight vectors expressing GAA under the control of a liver-specific promoter to the portal vein of Gaa–/– mice. Several important findings were highlighted in this study. In a previous study, Raben et al. used transgenic Gaa–/– mice that overexpress liver-specific hGAA and observed that cross-correction of muscles from liver-produced GAA required 10- to 12-fold normal GAA expression levels in the liver. Therefore, Cresawn et al. optimized liver-hGAA expression from rAAV vectors by testing three liver-specific promotes and two rAAV serotypes. The authors found that the duck hepatitis B viral core (DHBV) promoter yields up to eight times more liver-hGAA expression than more commonly used liver-specific promoters (albumin and human alpha–1-antitrypsin promoters). Additionally, rAAV serotype 8 vectors resulted in 2.8-fold higher levels of liver-hGAA expression that rAAV5 vectors. Using the rAAV8 vector with the DHBV promoter, an average of 26-fold over normal hGAA levels were observed in the livers of Gaa–/– mice 16 weeks after portal vein delivery. Based on previous observation of antibody formation to the liver-secreted hGAA protein, the authors investigated the extent and inhibitory effect of antibody formation in Gaa–/– mice after rAAV8-hGAA portal vein delivery. Results revealed anti-GAA antibody levels of 100-fold over untreated controls. More importantly, one group of mice was pre-treated with rhGAA protein at 1 day of age rendering them partially or completely immune tolerant. Comparison of GAA activities in the distal tissues revealed significantly higher levels of activity in the immune-tolerant mice compared to the immune-responsive mice in the heart, diaphragm, and all examined hind-limb muscles (gastrocnemius, quadriceps, tibialis anterior, and soleus). Additionally, the high levels of GAA expression in cross-corrected tissues of immune-tolerant mice only had clearance of glycogen and partial restoration of muscle contractile function. This observation of inhibition of cross-correction by anti-GAA antibodies was also observed earlier by Xu et al. (2004) where adenoviral-mediated delivery to an immunodeficient Gaa–/– mouse model resulted in significantly higher levels of GAA expression, histological, and functional correction than in the immunocompetent Gaa–/– mouse model.

Peripheral intravenous delivery could essentially achieve the same vector distribution as portal vein delivery where vector genomes are primarily detected in the liver and secondarily in other tissues depending on the promoter and serotype used. Unlike portal vein delivery, tissues (other than the liver) from animals treated intravenously may be corrected by direct transduction in addition to by uptake of liver-secreted protein. Peripheral IV delivery would be preferred over portal vein delivery for clinical safety concerns. However, the antibody-mediated inhibition of circulating GAA remains a concern for both portal vein and intravenous vector delivery approaches. In effort to understand the full capacity of intravenously delivered rAAV-hGAA vectors to correct the pathologies associated with GSDII, Mah et al. (2004) delivered rAAV-hGAA vectors intravenously to 1-day-old Gaa–/– mice and evaluated long-term correction 1 year after treatment. Because the immune system is not fully developed in mice until after birth, the immune system can recognize a foreign protein, such as GAA, as a self-protein and fail to elicit an immune response. Using this model, the authors observed 81% of normal levels of GAA activity in the heart 1 year post-injection. Achieving such high levels of sustained activity in the heart from a single intravenous injection was promising and suggested that direct intracardiac injections will not be necessary for correction of the cardiac phenotype as previously thought. This is, of course, dependent on the ability to treat patients that are immune-tolerant either naturally (due to residual levels of GAA production) or induced by immune-suppressive drugs.

The use of immune-suppressive drugs may be avoidable even in null patients, and studies to understand this are currently underway by several groups. Several theories have been proposed to explain the variations in antibody response among similar diseases models treated with rAAV vectors. Much of the work in this area has been performed by groups studying rAAV-mediated treatment for hemophilia. However, the observations remain applicable for LSDs. Several groups have reported that the ability to make a null animal tolerant to a foreign transgene product is dependent on the target tissue. Specifically, that the liver-directed delivery is more tolerizing than muscle-directed delivery. Ge et al. (2001) delivered rAAV vectors expressing human factor IX (hFIX) to the liver or muscle of immunocompetent mice at identical doses. While all of the intramuscular-treated mice developed a robust humoral immune response blocking hFIX expression in the serum, none of the intrahepatic-treated mice elicited an immune response coinciding with significant levels of serum hFIX. Using the same hemophilia disease model, Mingozzi et al. (2003) induced immune tolerance in immunocompetent mice by intrahepatic administration of rAAV-delivered hFIX. Tolerance of hFIX was challenged by administration of recombinant hFIX protein in adjuvant after pre-treatment with intrahepatic delivery of either rAAV-hFIX or rAAV-GFP. Mice pre-treated with rAAV-hFIX developed no anti-hFIX antibodies and had reduced in vitro T cell responses. However, mice pre-treated with rAAV-GFP developed anti-FIX antibodies 14 days after injection of rhFIX along with significantly higher in vitro T cell responses. Ziegler et al. (2004) reported similar findings in the Fabry mouse model. This group found that intravenous delivery of rAAV2 vectors expressing α-galactosidase under control of a liver-specific promoter did not result in an antibody response and, furthermore, induced tolerance to subsequent challenge with recombinant α-galacotosidase enzyme. However, when this experiment was performed using the ubiquitous CMV promoter, antibodies to the transgene product were detected at levels of 100-fold over untreated controls.

As progress is being made towards efficacy of rAAV-mediated gene therapy and interest in clinical trials is progressing, the need to fully understand the immunological impact of rAAV-mediated gene therapy is increasing. While there have been few reports of toxicity associated with the rAAV vector itself, there is the potential for an immune response to the encoded transgene product. The likelihood of an rAAV-delivered transgene eliciting an inhibitory immune response depends on several factors including the genetic background of the host, amount of protein secreted by transduced cells, presence or absence of residual mutant protein (Brooks, 1999), and the route of vector administration (Brockstedt et al., 1999; Cordier et al., 2001; Mingozzi et al., 2003). Because correction of many LSDs relies on the survival of circulating protein, these studies will have great impact in the LSD field.

10.9. Gene therapy for CNS pathologies in LSDs

Unlike GSDII, the major manifestation of many LSDs is neurological including mental retardation and seizures. Additionally, correction of other non-neurological LSDs, that otherwise result in early death, may require neurological intervention later in life. While gene therapy offers advantages to ERT for CNS correction (i.e., only single injection required for sustained expression), the idea of CNS delivery from systemically administered vector is limited by the confines of the blood–brain barrier. This results in the need for direct intracranial injections or pharmacological intervention to increase permeability. The BBB is a continuous layer of endothelial cells that plays a critical role in maintaining CNS homeostasis by protecting the CNS from the compositional fluctuations that occur in the blood. (Gloor et al., 2001). This physical separation of the CNS from the rest of the body prevents many therapeutic reagents including low molecular weight pharmaceuticals, purified enzyme, and gene therapy vectors from reaching the brain. One method of overcoming this barrier is the use of mannitol, a known BBB interruptive reagent. Intravenous infusion of 25% mannitol reduces intracranial pressure in patients with traumatic brain disease by pulling fluid from the CNS and temporarily increasing vascular osmotic pressure (Fu et al., 2003). Clinical studies evaluating the use of mannitol to open the BBB and enhance entrance of chemotherapeutic agents have resulted in improved survival of brain cancer patients. A variety of pre-clinical studies are underway evaluating the use of mannitol to enhance delivery of wide range of substances including antibodies, enzymes, and viral vectors (Fu et al., 2003). For example, Fu et al. (2003) delivered AAV2-GFP intravenously (via tail vein injection) to adult mice following infusion of 25% mannitol. This resulted in global distribution of GFP in the brain and spinal cord compared to background GFP levels in the CNS of mice injected with vector alone or in the presence of only 12.5% mannitol. It should be noted, however, that this vector was generated from the more recently developed self-complementary rAAV vector (McCarty et al., 2001), which has a higher transduction efficiency than the traditionally used single-stranded DNA rAAV vectors. The authors did not see any significant increase in GFP levels in the CNS or spinal cord when single-stranded rAAV vectors were administered following 25% mannitol infusion (Fu et al., 2003).

Another important study examining approaches to overcome the BBB was performed by Urayama et al. (2004). This study examined age-dependent uptake of systemically administered enzyme in the brain of MPS VII mice. This interest was based on a previous finding that systemically administered β-glucoronidase (GUSB) enzyme was able to correct the neurological phenotype in MPS VII mice, if administered prior to 2 weeks of age. In the follow-up study, Urayama et al. (2004) found that radio-labeled, phosphorylated enzyme, transported across the BBB after intravenous delivery in 2-day-old mice. The influx rate decreased with age. When enzyme was administered at 7 weeks there was no significant transport of enzyme to the brain. This process was inhibited by infusion of mannose 6-phosphate in a dose-dependent manner and non-phosphorylated enzyme entered the brain at a significantly slower rate. These results suggest that there is an age-dependent downregulation of mannose 6-phosphate mediated transport of phosphorylated enzyme into the brain. Based on earlier studies by Nissley et al., this observation is most likely due to developmentally regulated changes in BBB permeability and not the downregulation of the M6P receptor. Nissley et al. compared protein levels at 20 days fetal with 20 days postnatal. Results showed an insignificant decrease in M6P levels in the brain of <1% compared to 20% decrease in heart and skeletal muscle Nissley et al. (1993).

One mechanism to avoid the limitations of intravenous delivery for CNS correction is retrograde transport. Retrograde transport has been reported using intramuscular or spinal cord injection of AAV vectors by several groups (Lu et al., 2003; Burger et al., 2004). The principle of retrograde transport is based on the idea that intramuscular or spinal cord injections resulting in transduced motor neurons will transport vector to the cell body, uncoat, and then translocate to the nucleus of the neuron (Burger et al., 2004). For example, Burger et al. observed significant levels of GFP in the hippocampus, motor cortex, and substantia nigra after spinal cord injections of AAV2 and AAV5-GFP in mice.

Retrograde transport studies, the use of reagents such as mannitol, and furthering our understanding of the critical parameters (such as age) in increasing gene delivery to the brain from a systemic delivery method are important areas of research to improve the safety and feasibility of gene therapy-mediated correction of neurological disorders. An alternative way to circumvent these limitations is by direct intracranial delivery. Extensive studies have been done examining the tropism, transduction efficiency, and virus spread of various rAAV serotypes in the brain from intracranial delivery. Recombinant AAV vectors are good candidates for CNS gene therapy as they are able to transduce postmitotic cells. The first clinical trial involving neurosurgical delivery of AAV2 vectors has been initiated in patients with Canavan disease (Janson et al., 2002; Burger et al., 2004). Recombinant AAV2 vectors have been shown to tranduce the hippocampus, substantia niagra, striatum, piriform, and lateral cortex, olfactory tubercle, cerebellum, inferior colliculus, globus pallidus, basal forebrain, subthalmic nucleus, facial motor nucleus, and spinal cord. Transduction efficiencies vary among the different regions and correlate with the level of AAV2 co-receptor (FGFR) with the highest levels observed in the hippocampus, inferior colliculus, and piriform cortex followed by olfactory tubercle, followed by the striatum (Takami et al., 1998; Tenenbaum et al., 2004).

While most of the information concerning AAV transduction in the brain has been learned with AAV2 vectors, recent studies using alternative serotypes have revealed important differences. For example, AAV2 has been shown to transduce primarily neurons whereas AAV5 vectors transduce both neuronal and glial cells (Davidson et al., 2000). Burger et al. (2004) showed that AAV1 and 5 vectors exhibit higher transduction frequencies than AAV2 vectors in all regions injected including the hippocampus, substantia nigra, striatum, globus pallidus, and spinal cord. Along with higher transduction efficiency, AAV1 vectors have also been shown to yield wider distribution in the striatum and entire midbrain than AAV2 vectors (Wang et al., 2002; Burger et al., 2004). Strategies to enhance virus spread include coadministration of heparin (to prevent the sequestration of heparin sulfate proteoglycan receptor-binding AAV2 vectors at the site of injection) and multiple injections in different regions of the brain.

Extensive studies on gene therapy-mediated correction of the CNS pathologies in LSDs have been reported using the MPS VII model and is reviewed by Ellinwood et al. (2004). MPS VII (Sly syndrome) results from a deficiency in the enzyme β-glucoronidase and has multiple manifestations including organomegaly, skeletal, cardiovascular, ocular, and neurological. With both the dog and mouse models of MPS VII, multiple methods of intracranial delivery have been reported in both adult and neonatal ages of the two animal models. Other examples of intracranial delivery for treatment of LSDs include MPS I, MPS IIIB, Krabbe, and metachromatic leukodystrophy (Table 10.1).

10.10. Conclusion

Gene therapy for the treatment of lysosomal storage diseases has made considerable progress in the past 10 years. Establishing proof that vector-delivered transgene product can restore biochemical activity to affected cells, cross-correct untransduced cells, and restore histological and functional phenotype, is the first obstacle and has already been demonstrated for a variety of LSDs. The issues that lie ahead as scientists look towards clinical trials include the following: (1) Vector and transgene product toxicity. Specifically, what is the potential for cell-mediated and/or antibody-mediated immune response to the vector and/or transgene product, and how does this response affect efficacy and patient safety. (2) Further developing delivery methods to systemically correct the CNS pathologies. (3) Optimizing gene regulatory elements, engineered secretion signals, vectors, and serotypes to maximize transgene expression and protein secretion to take full advantage of the secretion-reuptake pathway of lysosomal enzymes. While the frequency of individual lysosomal storage disorders is rare, the aggregate incidence of 1 in 7000 is not. With further progress at the bench and future success with current gene therapy clinical trials, the studies presented in this chapter can hopefully be applied to the many patients affected with LSDs and this large group of disorders can join other diseases such as cancer and cystic fibrosis as part of a new age in genetics and medicine.

References

Amalfitano, A., Vie-Wylie, A. J., Hu, H., Dawson, T. L., Raben, N., Plotz, P. and Chen, Y. T. (1999). Systemic correction of the muscle disorder glycogen storage disease type II after hepatic targeting of a modified adenovirus vector encoding human acid-alpha-glucosidase. Proc. Natl. Acad. Sci. USA 96, 8861–8866.

Asano, N., Ishii, S., Kizu, H., Ikeda, K., Yasuda, K., Kato, A., Martin, O. R. and Fan, J. Q. (2000). In vitro inhibition and intracellular enhancement of lysosomal alpha-galactosidase A activity in Fabry lymphoblasts by 1-deoxygalactonojirimycin and its derivatives. Eur. J. Biochem. 267, 4179–4186.

Bijvoet, A. G., Kroos, M. A., Pieper, F. R., Van, d., V, De Boer, H. A., Van der Ploeg, A. T., Verbeet, M. P. and Reuser, A. J. (1998). Recombinant human acid alpha-glucosidase: High level production in mouse milk, biochemical characteristics, correction of enzyme deficiency in GSDII KO mice. Hum. Mol. Genet. 7, 1815–1824.

Bodamer, O. A., Halliday, D. and Leonard, J. V. (2000). The effects of l-alanine supplementation in late-onset glycogen storage disease type II. Neurology 55, 710–712.

Bodamer, O. A., Leonard, J. V. and Halliday, D. (1997). Dietary treatment in late-onset acid maltase deficiency. Eur. J. Pediatr. 156(Suppl. 1), S39–S42.

Bourgoin, C., Emiliani, C., Kremer, E. J., Gelot, A., Tancini, B., Gravel, R. A., Drugan, C., Orlacchio, A., Poenaru, L. and Caillaud, C. (2003). Widespread distribution of beta-hexosaminidase activity in the brain of a Sandhoff mouse model after coinjection of adenoviral vector and mannitol. Gene Ther. 10, 1841–1849.

Brockstedt, D. G., Podsakoff, G. M., Fong, L., Kurtzman, G., Mueller-Ruchholtz, W. and Engleman, E. G. (1999). Induction of immunity to antigens expressed by recombinant adeno-associated virus depends on the route of administration. Clin. Immunol. 92, 67–75.

Brooks, D. A. (1999). Immune response to enzyme replacement therapy in lysosomal storage disorder patients and animal models. Mol. Genet. Metab. 68, 268–275.

Burger, C., Gorbatyuk, O. S., Velardo, M. J., Peden, C. S., Williams, P., Zolotukhin, S., Reier, P. J., Mandel, R. J. and Muzyczka, N. (2004). Recombinant AAV viral vectors pseudotyped with viral capsids from serotypes 1, 2, and 5 display differential efficiency and cell tropism after delivery to different regions of the central nervous system. Mol. Ther. 10, 302–317.

Consiglio, A., Quattrini, A., Martino, S., Bensadoun, J. C., Dolcetta, D., Trojani, A., Benaglia, G., Marchesini, S., Cestari, V., Oliverio, A., Bordignon, C. and Naldini, L. (2001). In vivo gene therapy of metachromatic leukodystrophy by lentiviral vectors: Correction of neuropathology and protection against learning impairments in affected mice. Nat. Med. 7, 310–316.

Cordier, L., Gao, G. P., Hack, A. A., McNally, E. M., Wilson, J. M., Chirmule, N. and Sweeney, H. L. (2001). Muscle-specific promoters may be necessary for adeno-associated virus-mediated gene transfer in the treatment of muscular dystrophies. Hum. Gene Ther. 12, 205–215.

Cresawn, K. O., Fraites, T. J., Wasserfall, C., Atkinson, M., Lewis, M. A., Porvasnik, S. L., Mah, C. and Byrne, B. (2005). Hum. Gene Ther. 16(1), 68–80.

Davidson, B. L., Stein, C. S., Heth, J. A., Martins, I., Kotin, R. M., Derksen, T. A., Zabner, J., Ghodsi, A. and Chiorini, J. A. (2000). Recombinant adeno-associated virus type 2, 4, and 5 vectors: Transduction of variant cell types and regions in the mammalian central nervous system. Proc. Natl. Acad. Sci. USA 97, 3428–3432.

Desmaris, N., Verot, L., Puech, J. P., Caillaud, C., Vanier, M. T. and Heard, J. M. (2004). Prevention of neuropathology in the mouse model of Hurler syndrome. Ann. Neurol. 56, 68–76.

Ding, E. Y., Hodges, B. L., Hu, H., McVie-Wylie, A. J., Serra, D., Migone, F. K., Pressley, D., Chen, Y. T. and Amalfitano, A. (2001). Long-term efficacy after [E1-, polymerase-] adenovirus-mediated transfer of human acid-alpha-glucosidase gene into glycogen storage disease type II knockout mice. Hum. Gene Ther. 12, 955–965.

Ding, E., Hu, H., Hodges, B. L., Migone, F., Serra, D., Xu, F., Chen, Y. T. and Amalfitano, A. (2002). Efficacy of gene therapy for a prototypical lysosomal storage disease (GSD-II) is critically dependent on vector dose, transgene promoter, and the tissues targeted for vector transduction. Mol. Ther. *5*, 436–446.

Du, H., Heur, M., Witte, D. P., Ameis, D. and Grabowski, G. A. (2002). Lysosomal acid lipase deficiency: Correction of lipid storage by adenovirus-mediated gene transfer in mice. Hum. Gene Ther. *13*, 1361–1372.

Dunbar, C. E., Kohn, D. B., Schiffmann, R., Barton, N. W., Nolta, J. A., Esplin, J. A., Pensiero, M., Long, Z., Lockey, C., Emmons, R. V., Csik, S., Leitman, S., Krebs, C. B., Carter, C., Brady, R. O. and Karlsson, S. (1998). Retroviral transfer of the glucocerebrosidase gene into CD34+ cells from patients with Gaucher disease: In vivo detection of transduced cells without myeloablation. Hum. Gene Ther. *9*, 2629–2640.

Ellinwood, N. M., Vite, C. H. and Haskins, M. E. (2004). Gene therapy for lysosomal storage diseases: The lessons and promise of animal models. J. Gene Med. *6*, 481–506.

Eng, C. M., Banikazemi, M., Gordon, R. E., Goldman, M., Phelps, R., Kim, L., Gass, A., Winston, J., Dikman, S., Fallon, J. T., Brodie, S., Stacy, C. B., Mehta, D., Parsons, R., Norton, K., O'Callaghan, M. and Desnick, R. J. (2001). A phase 1/2 clinical trial of enzyme replacement in fabry disease: Pharmacokinetic, substrate clearance, and safety studies. Am. J. Hum. Genet. *68*, 711–722.

Eng, C. M., Guffon, N., Wilcox, W. R., Germain, D. P., Lee, P., Waldek, S., Caplan, L., Linthorst, G. E. and Desnick, R. J. (2001). Safety and efficacy of recombinant human alpha-galactosidase A—replacement therapy in Fabry's disease. N. Engl. J. Med. *345*, 9–16.

Fan, J. Q. (2003). A contradictory treatment for lysosomal storage disorders: Inhibitors enhance mutant enzyme activity. Trends Pharmacol. Sci. *24*, 355–360.

Fraites, T. J., Jr., Schleissing, M. R., Shanely, R. A., Walter, G. A., Cloutier, D. A., Zolotukhin, I., Pauly, D. F., Raben, N., Plotz, P. H., Powers, S. K., Kessler, P. D. and Byrne, B. J. (2002). Correction of the enzymatic and functional deficits in a model of Pompe disease using adeno-associated virus vectors. Mol. Ther. *5*, 571–578.

Fu, H., Muenzer, J., Samulski, R. J., Breese, G., Sifford, J., Zeng, X. and McCarty, D. M. (2003). Self-complementary adeno-associated virus serotype 2 vector: Global distribution and broad dispersion of AAV-mediated transgene expression in mouse brain. Mol. Ther. *8*, 911–917.

Fu, H., Samulski, R. J., McCown, T. J., Picornell, Y. J., Fletcher, D. and Muenzer, J. (2002). Neurological correction of lysosomal storage in a mucopolysaccharidosis IIIB mouse model by adeno-associated virus-mediated gene delivery. Mol. Ther. *5*, 42–49.

Gao, G. P., Alvira, M. R., Wang, L., Calcedo, R., Johnston, J. and Wilson, J. M. (2002). Novel adeno-associated viruses from rhesus monkeys as vectors for human gene therapy. Proc. Natl. Acad. Sci. USA 99, 11854–11859.

Ge, Y., Powell, S., Van Roey, M. and McArthur, J. G. (2001). Factors influencing the development of an anti-factor IX (FIX) immune response following administration of adeno-associated virus-FIX. Blood 97, 3733–3737.

Gloor, S. M., Wachtel, M., Bolliger, M. F., Ishihara, H., Landmann, R. and Frei, K. (2001). Molecular and cellular permeability control at the blood–brain barrier. Brain Res. Brain Res. Rev. 36, 258–264.

Griffey, M., Bible, E., Vogler, C., Levy, B., Gupta, P., Cooper, J. and Sands, M. S. (2004). Adeno-associated virus 2-mediated gene therapy decreases autofluorescent storage material and increases brain mass in a murine model of infantile neuronal ceroid lipofuscinosis. Neurobiol. Dis. 16, 360–369.

Guidotti, J. E., Mignon, A., Haase, G., Caillaud, C., McDonell, N., Kahn, A. and Poenaru, L. (1999). Adenoviral gene therapy of the Tay-Sachs disease in hexosaminidase A-deficient knock-out mice. Hum. Mol. Genet. 8, 831–838.

Harmatz, P., Whitley, C. B., Waber, L., Pais, R., Steiner, R., Plecko, B., Kaplan, P., Simon, J., Butensky, E. and Hopwood, J. J. (2004). Enzyme replacement therapy in mucopolysaccharidosis VI (Maroteaux-Lamy syndrome). J. Pediatr. 144, 574–580.

Hartung, S. D., Frandsen, J. L., Pan, D., Koniar, B. L., Graupman, P., Gunther, R., Low, W. C., Whitley, C. B. and McIvor, R. S. (2004). Correction of metabolic, craniofacial, and neurologic abnormalities in MPS I mice treated at birth with adeno-associated virus vector transducing the human alpha-L-iduronidase gene. Mol. Ther. 9, 866–875.

Haskell, R. E., Hughes, S. M., Chiorini, J. A., Alisky, J. M. and Davidson, B. L. (2003). Viral-mediated delivery of the late-infantile neuronal ceroid lipofuscinosis gene, TPP-I to the mouse central nervous system. Gene Ther. 10, 34–42.

Hirschhorn, R. and Reuser, A. J. J. (2000). In: Metabolic Basis of Inherited Disease (Scriver, C. R., Beaudet, A. L., Sly, W. S. and Valle, D., eds.). Macmillan, New York, pp. 3389–3420.

Hong, Y. B., Kim, E. Y., Yoo, H. W. and Jung, S. C. (2004). Feasibility of gene therapy in Gaucher disease using an adeno-associated virus vector. J. Hum. Genet. 49(10), 836–843.

Janson, C., McPhee, S., Bilaniuk, L., Haselgrove, J., Testaiuti, M., Freese, A., Wang, D. J., Shera, D., Hurh, P., Rupin, J., Saslow, E., Goldfarb, O.,

Goldberg, M., Larijani, G., Sharrar, W., Liouterman, L., Camp, A., Kolodny, E., Samulski, J. and Leone, P. (2002). Clinical protocol Gene therapy of Canavan disease: AAV-2 vector for neurosurgical delivery of aspartoacylase gene (ASPA) to the human brain. Hum. Gene Ther. *13*, 1391–1412.

Jin, H. K. and Schuchman, E. H. (2003). Ex vivo gene therapy using bone marrow-derived cells: Combined effects of intracerebral and intravenous transplantation in a mouse model of Niemann-Pick disease. Mol. Ther. *8*, 876–885.

Jung, S. C., Han, I. P., Limaye, A., Xu, R., Gelderman, M. P., Zerfas, P., Tirumalai, K., Murray, G. J., During, M. J., Brady, R. O. and Qasba, P. (2001). Adeno-associated viral vector-mediated gene transfer results in long-term enzymatic and functional correction in multiple organs of Fabry mice. Proc. Natl. Acad. Sci. USA *98*, 2676–2681.

Kakkis, E. D., Schuchman, E., He, X., Wan, Q., Kania, S., Wiemelt, S., Hasson, C. W., O'Malley, T., Weil, M. A., Aguirre, G. A., Brown, D. E. and Haskins, M. E. (2001). Enzyme replacement therapy in feline mucopolysaccharidosis I. Mol. Genet. Metab. *72*, 199–208.

Kim, E. Y., Hong, Y. B., Lai, Z., Kim, H. J., Cho, Y. H., Brady, R. O. and Jung, S. C. (2004). Expression and secretion of human glucocerebrosidase mediated by recombinant lentivirus vectors in vitro and in vivo: Implications for gene therapy of Gaucher disease. Biochem. Biophys. Res. Commun. *318*, 381–390.

Lacorazza, H. D., Flax, J. D., Snyder, E. Y. and Jendoubi, M. (1996). Expression of human beta-hexosaminidase alpha-subunit gene (the gene defect of Tay-Sachs disease) in mouse brains upon engraftment of transduced progenitor cells. Nat. Med. *2*, 424–429.

Laine, M., Ahtiainen, L., Rapola, J., Richter, J. and Jalanko, A. (2004). Bone marrow transplantation in young aspartylglucosaminuria mice: Improved clearance of lysosomal storage in brain by using wild type as compared to heterozygote donors. Bone Marrow Transplant. *34*, 1001–1003.

Leimig, T., Mann, L., Martin, M. P., Bonten, E., Persons, D., Knowles, J., Allay, J. A., Cunningham, J., Nienhuis, A. W., Smeyne, R. and d'Azzo, A. (2002). Functional amelioration of murine galactosialidosis by genetically modified bone marrow hematopoietic progenitor cells. Blood *99*, 3169–3178.

Li, C., Ziegler, R. J., Cherry, M., Lukason, M., Desnick, R. J., Yew, N. S. and Cheng, S. H. (2002). Adenovirus-transduced lung as a portal for delivering alpha-galactosidase A into systemic circulation for Fabry disease. Mol. Ther. *5*, 745–754.

Lin, C. Y., Ho, C. H., Hsieh, Y. H. and Kikuchi, T. (2002). Adeno-associated virus-mediated transfer of human acid maltase gene results in a transient reduction of glycogen accumulation in muscle of Japanese quail with acid maltase deficiency. Gene Ther. 9, 554–563.

Liu, C., Dunigan, J. T., Watkins, S. C., Bahnson, A. B. and Barranger, J. A. (1998). Long-term expression, systemic delivery, and macrophage uptake of recombinant human glucocerebrosidase in mice transplanted with genetically modified primary myoblasts. Hum. Gene Ther. 9, 2375–2384.

Lu, Y. Y., Wang, L. J., Muramatsu, S., Ikeguchi, K., Fujimoto, K., Okada, T., Mizukami, H., Matsushita, T., Hanazono, Y., Kume, A., Nagatsu, T., Ozawa, K. and Nakano, I. (2003). Intramuscular injection of AAV-GDNF results in sustained expression of transgenic GDNF, and its delivery to spinal motoneurons by retrograde transport. Neurosci. Res. 45, 33–40.

Lutzko, C., Kruth, S., Abrams-Ogg, A. C., Lau, K., Li, L., Clark, B. R., Ruedy, C., Nanji, S., Foster, R., Kohn, D., Shull, R. and Dube, I. D. (1999). Genetically corrected autologous stem cells engraft, but host immune responses limit their utility in canine alpha-L-iduronidase deficiency. Blood 93, 1895–1905.

Mah, C., Cresawn, K. O., Fraites, T. J., Lewis, M. A., Zolotukhin, I. and Byrne, B. (in press). Sustained correction of glycogen storage disease type II using adeno-associated virus serotype I vectors. Gene Ther.

Mah, C., Fraites, T. J., Jr., Cresawn, K. O., Zolotukhin, I., Lewis, M. A. and Byrne, B. J. (2004). A new method for recombinant adeno-associated virus vector delivery to murine diaphragm. Mol. Ther. 9, 458–463.

Marshall, J., McEachern, K. A., Kyros, J. A., Nietupski, J. B., Budzinski, T., Ziegler, R. J., Yew, N. S., Sullivan, J., Scaria, A., van Rooijen, N., Barranger, J. A. and Cheng, S. H. (2002). Demonstration of feasibility of in vivo gene therapy for Gaucher disease using a chemically induced mouse model. Mol. Ther. 6, 179–189.

Martiniuk, F., Chen, A., Mack, A., Donnabella, V., Slonim, A., Bulone, L., Arvanitopoulos, E., Raben, N., Plotz, P. and Rom, W. N. (2002). Helios gene gun particle delivery for therapy of acid maltase deficiency. DNA Cell Biol. 21, 717–725.

Martin-Touaux, E., Puech, J. P., Chateau, D., Emiliani, C., Kremer, E. J., Raben, N., Tancini, B., Orlacchio, A., Kahn, A. and Poenaru, L. (2002). Muscle as a putative producer of acid alpha-glucosidase for glycogenosis type II gene therapy. Hum. Mol. Genet. 11, 1637–1645.

Matzner, U., Habetha, M. and Gieselmann, V. (2000a). Retrovirally expressed human arylsulfatase A corrects the metabolic defect of arylsulfatase A-deficient mouse cells. Gene Ther. 7, 805–812.

Matzner, U., Harzer, K., Learish, R. D., Barranger, J. A. and Gieselmann, V. (2000b). Long-term expression and transfer of arylsulfatase A into brain of arylsulfatase A-deficient mice transplanted with bone marrow expressing the arylsulfatase A cDNA from a retroviral vector. Gene Ther. 7, 1250–1257.

Matzner, U., Hartmann, D., Lullmann-Rauch, R., Coenen, R., Rothert, F., Mansson, J. E., Fredman, P., D'Hooge, R., De Deyn, P. P. and Gieselmann, V. (2002). Bone marrow stem cell-based gene transfer in a mouse model for metachromatic leukodystrophy: Effects on visceral and nervous system disease manifestations. Gene Ther. 9, 53–63.

McCarty, D. M., Monahan, P. E. and Samulski, R. J. (2001). Self-complementary recombinant adeno-associated virus (scAAV) vectors promote efficient transduction independently of DNA synthesis. Gene Ther. 8, 1248–1254.

Meertens, L., Zhao, Y., Rosic-Kablar, S., Li, L., Chan, K., Dobson, H., Gartley, C., Lutzko, C., Hopwood, J., Kohn, D., Kruth, S., Hough, M. R. and Dube, I. D. (2002). In utero injection of alpha-L-iduronidase-carrying retrovirus in canine mucopolysaccharidosis type I: Infection of multiple tissues and neonatal gene expression. Hum. Gene Ther. 13, 1809–1820.

Mingozzi, F., Liu, Y. L., Dobrzynski, E., Kaufhold, A., Liu, J. H., Wang, Y., Arruda, V. R., High, K. A. and Herzog, R. W. (2003). Induction of immune tolerance to coagulation factor IX antigen by in vivo hepatic gene transfer. J. Clin. Invest. 111, 1347–1356.

Mingozzi, F., Schuttrumpf, J., Arruda, V. R., Liu, Y., Liu, Y. L., High, K. A., Xiao, W. and Herzog, R. W. (2002). Improved hepatic gene transfer by using an adeno-associated virus serotype 5 vector. J. Virol. 76, 10497–10502.

Miranda, S. R., Erlich, S., Friedrich, V. L., Jr., Gatt, S. and Schuchman, E. H. (2000). Hematopoietic stem cell gene therapy leads to marked visceral organ improvements and a delayed onset of neurological abnormalities in the acid sphingomyelinase deficient mouse model of Niemann-Pick disease. Gene Ther. 7, 1768–1776.

Muenzer, J., Lamsa, J. C., Garcia, A., Dacosta, J., Garcia, J. and Treco, D. A. (2002). Enzyme replacement therapy in mucopolysaccharidosis type II (Hunter syndrome): A preliminary report. Acta Paediatr. Suppl. 91, 98–99.

Nicolino, M. P., Puech, J. P., Kremer, E. J., Reuser, A. J., Mbebi, C., Verdiere-Sahuque, M., Kahn, A. and Poenaru, L. (1998). Adenovirus-mediated transfer of the acid alpha-glucosidase gene into fibroblasts, myoblasts and myotubes from patients with glycogen storage disease type II leads to high level expression of enzyme and corrects glycogen accumulation. Hum. Mol. Genet. *7*, 1695–1702.

Nissley, P., Kiess, W. and Sklar, M. (1993). Developmental expression of the IGF-II/mannose 6-phosphate receptor. Mol. Reprod. Dev. *35*, 408–413.

Park, J., Murray, G. J., Limaye, A., Quirk, J. M., Gelderman, M. P., Brady, R. O. and Qasba, P. (2003). Long-term correction of globotriaosylceramide storage in Fabry mice by recombinant adeno-associated virus-mediated gene transfer. Proc. Natl. Acad. Sci. USA *100*, 3450–3454.

Pauly, D. F., Fraites, T. J., Toma, C., Bayes, H. S., Huie, M. L., Hirschhorn, R., Plotz, P. H., Raben, N., Kessler, P. D. and Byrne, B. J. (2001). Intercellular transfer of the virally derived precursor form of acid alpha-glucosidase corrects the enzyme deficiency in inherited cardioskeletal myopathy Pompe disease. Hum. Gene Ther. *12(5)*, 527–538.

Pauly, D. F., Johns, D. C., Matelis, L. A., Lawrence, J. H., Byrne, B. J. and Kessler, P. D. (1998). Complete correction of acid alpha-glucosidase deficiency in Pompe disease fibroblasts in vitro, and lysosomally targeted expression in neonatal rat cardiac and skeletal muscle. Gene Ther. *5*, 473–480.

Raben, N., Nagaraju, K., Lee, E., Kessler, P., Byrne, B., Lee, L., La Marca, M., King, C., Ward, J., Sauer, B. and Plotz, P. (1998). Targeted disruption of the acid alpha-glucosidase gene in mice causes an illness with critical features of both infantile and adult human glycogen storage disease type II. J. Biol. Chem. *273*, 19086–19092.

Raben, N., Nagaraju, K., Lee, E. and Plotz, P. (2000). Modulation of disease severity in mice with targeted disruption of the acid alpha-glucosidase gene. Neuromuscul. Disord. *10*, 283–291.

Raben, N., Plotz, P. and Byrne, B. J. (2002). Acid alpha-glucosidase deficiency (glycogenosis type II, Pompe disease). Curr. Mol. Med. *2*, 145–166.

Raper, S. E., Chirmule, N., Lee, F. S., Wivel, N. A., Bagg, A., Gao, G. P., Wilson, J. M. and Batshaw, M. L. (2003). Fatal systemic inflammatory response syndrome in a ornithine transcarbamylase deficient patient following adenoviral gene transfer. Mol. Genet. Metab. *80*, 148–158.

Reuser, A. J., Kroos, M. A., Hermans, M. M., Bijvoet, A. G., Verbeet, M. P., Van Diggelen, O. P., Kleijer, W. J. and Van der Ploeg, A. T. (1995). Glycogenosis type II (acid maltase deficiency). Muscle Nerve *3*, S61–S69.

Robbins, P. D., Tahara, H. and Ghivizzani, S. C. (1998). Viral vectors for gene therapy. Trends Biotechnol. *16*, 35–40.

Rucker, M., Fraites, T. J., Jr., Porvasnik, S. L., Lewis, M. A., Zolotukhin, I., Cloutier, D. A. and Byrne, B. J. (2004). Rescue of enzyme deficiency in embryonic diaphragm in a mouse model of metabolic myopathy: Pompe disease. Development *131*, 3007–3019.

Sawkar, A. R., Cheng, W. C., Beutler, E., Wong, C. H., Balch, W. E. and Kelly, J. W. (2002). Chemical chaperones increase the cellular activity of N370S beta-glucosidase: A therapeutic strategy for Gaucher disease. Proc. Natl. Acad. Sci. USA 99, 15428–15433.

Shen, J. S., Watabe, K., Ohashi, T. and Eto, Y. (2001). Intraventricular administration of recombinant adenovirus to neonatal twitcher mouse leads to clinicopathological improvements. Gene Ther. 8, 1081–1087.

Simonaro, C. M., Haskins, M. E., Abkowitz, J. L., Brooks, D. A., Hopwood, J. J., Zhang, J. and Schuchman, E. H. (1999). Autologous transplantation of retrovirally transduced bone marrow or neonatal blood cells into cats can lead to long-term engraftment in the absence of myeloablation. Gene Ther. 6, 107–113.

Sun, B. D., Chen, Y. T., Bird, A., Amalfitano, A. and Koeberl, D. D. (2003). Long-term correction of glycogen storage disease type II with a hybrid Ad-AAV vector. Mol. Ther. 7, 193–201.

Takahashi, H., Hirai, Y., Migita, M., Seino, Y., Fukuda, Y., Sakuraba, H., Kase, R., Kobayashi, T., Hashimoto, Y. and Shimada, T. (2002). Long-term systemic therapy of Fabry disease in a knockout mouse by adeno-associated virus-mediated muscle-directed gene transfer. Proc. Natl. Acad. Sci. USA 99, 13777–13782.

Takami, K., Matsuo, A., Terai, K., Walker, D. G., McGeer, E. G. and McGeer, P. L. (1998). Fibroblast growth factor receptor-1 expression in the cortex and hippocampus in Alzheimer's disease. Brain Res. 802, 89–97.

Takaura, N., Yagi, T., Maeda, M., Nanba, E., Oshima, A., Suzuki, Y., Yamano, T. and Tanaka, A. (2003). Attenuation of ganglioside GM1 accumulation in the brain of GM1 gangliosidosis mice by neonatal intravenous gene transfer. Gene Ther. 10, 1487–1493.

Takenaka, T., Murray, G. J., Qin, G., Quirk, J. M., Ohshima, T., Qasba, P., Clark, K., Kulkarni, A. B., Brady, R. O. and Medin, J. A. (2000). Long-term enzyme correction and lipid reduction in multiple organs of primary and secondary transplanted Fabry mice receiving transduced bone marrow cells. Proc. Natl. Acad. Sci. USA 97, 7515–7520.

Tenenbaum, L., Chtarto, A., Lehtonen, E., Velu, T., Brotchi, J. and Levivier, M. (2004). Recombinant AAV-mediated gene delivery to the central nervous system. J. Gene Med. 6(Suppl. 1), S212–S222.

Tomanin, R., Friso, A., Alba, S., Piller, P. E., Mennuni, C., La Monica, N., Hortelano, G., Zacchello, F. and Scarpa, M. (2002). Non-viral transfer approaches for the gene therapy of mucopolysaccharidosis type II (Hunter syndrome). Acta Paediatr. 91(Suppl.), 100–104.

Urayama, A., Grubb, J. H., Sly, W. S. and Banks, W. A. (2004). Developmentally regulated mannose 6-phosphate receptor-mediated transport of a lysosomal enzyme across the blood–brain barrier. Proc. Natl. Acad. Sci. USA *101*, 12658–12663.

Walvoort, H. C. (1983). Glycogen storage diseases in animals and their potential value as models of human disease. J. Inherit. Metab. Dis. *6*, 3–16.

Wang, L., Muramatsu, S., Lu, Y., Ikeguchi, K., Fujimoto, K., Okada, T., Mizukami, H., Hanazono, Y., Kume, A., Urano, F., Ichinose, H., Nagatsu, T., Nakano, I. and Ozawa, K. (2002). Delayed delivery of AAV-GDNF prevents nigral neurodegeneration and promotes functional recovery in a rat model of Parkinson's disease. Gene Ther. *9*, 381–389.

Weinreb, N. J., Charrow, J., Andersson, H. C., Kaplan, P., Kolodny, E. H., Mistry, P., Pastores, G., Rosenbloom, B. E., Scott, C. R., Wappner, R. S. and Zimran, A. (2002). Effectiveness of enzyme replacement therapy in 1028 patients with type 1 Gaucher disease after 2 to 5 years of treatment: A report from the Gaucher Registry. Am. J. Med. *113*, 112–119.

Wraith, J. E., Clarke, L. A., Beck, M., Kolodny, E. H., Pastores, G. M., Muenzer, J., Rapoport, D. M., Berger, K. I., Swiedler, S. J., Kakkis, E. D., Braakman, T., Chadbourne, E., Walton-Bowen, K. and Cox, G. F. (2004). Enzyme replacement therapy for mucopolysaccharidosis I: A randomized, double-blinded, placebo-controlled, multinational study of recombinant human alpha-L-iduronidase (laronidase). J. Pediatr. *144*, 581–588.

Xiao, W., Chirmule, N., Berta, S. C., McCullough, B., Gao, G. and Wilson, J. M. (1999). Gene therapy vectors based on adeno-associated virus type 1. J. Virol. *73*, 3994–4003.

Xu, F., Ding, E., Liao, S. X., Migone, F., Dai, J., Schneider, A., Serra, D., Chen, Y. T. and Amalfitano, A. (2004). Improved efficacy of gene therapy approaches for Pompe disease using a new, immune-deficient GSD-II mouse model. Gene Ther. *11(21)*, 1590–1598.

Yamaguchi, A., Katsuyama, K., Suzuki, K., Kosaka, K., Aoki, I. and Yamanaka, S. (2003). Plasmid-based gene transfer ameliorates visceral storage in a mouse model of Sandhoff disease. J. Mol. Med. *81*, 185–193.

Zaretsky, J. Z., Candotti, F., Boerkoel, C., Adams, E. M., Yewdell, J. W., Blaese, R. M. and Plotz, P. H. (1997). Retroviral transfer of acid alpha-glucosidase cDNA to enzyme-deficient myoblasts results in phenotypic spread of the genotypic correction by both secretion and fusion. Hum. Gene Ther. *8*, 1555–1563.

Zheng, Y., Ryazantsev, S., Ohmi, K., Zhao, H. Z., Rozengurt, N., Kohn, D. B. and Neufeld, E. F. (2004). Retrovirally transduced bone marrow has a therapeutic effect on brain in the mouse model of mucopolysaccharidosis IIIB. Mol. Genet. Metab. *82*, 286–295.

Ziegler, R. J., Lonning, S. M., Armentano, D., Li, C., Souza, D. W., Cherry, M., Ford, C., Barbon, C. M., Desnick, R. J., Gao, G., Wilson, J. M.,

Peluso, R., Godwin, S., Carter, B. J., Gregory, R. J., Wadsworth, S. C. and Cheng, S. H. (2004). AAV2 vector harboring a liver-restricted promoter facilitates sustained expression of therapeutic levels of alpha-galactosidase A and the induction of immune tolerance in Fabry mice. Mol. Ther. *9*, 231–240.

Ziegler, R. J., Yew, N. S., Li, C., Cherry, M., Berthelette, P., Romanczuk, H., Ioannou, Y. A., Zeidner, K. M., Desnick, R. J. and Cheng, S. H. (1999). Correction of enzymatic and lysosomal storage defects in Fabry mice by adenovirus-mediated gene transfer. Hum. Gene Ther. *10*, 1667–1682.

Zolotukhin, S., Potter, M., Zolotukhin, I., Sakai, Y., Loiler, S., Fraites, T. J., Jr., Chiodo, V. A., Phillipsberg, T., Muzyczka, N., Hauswirth, W. W., Flotte, T. R., Byrne, B. J. and Snyder, R. O. (2002). Production and purification of serotype 1, 2, and 5 recombinant adeno-associated viral vectors. Methods *28*, 158–167.

Index

A

AAV cellular receptors, 12, 20, 27
    αVβ5 integrin, 12, 20–21,
        90, 104, 169
    FGFR1, 20, 264
    heparan sulfate proteoglycan
        (HSPG), 12, 90, 169
    PDGFR, 21
    sialic acid, 21
AAV characterization,
    31, 36–37, 43
    AAV purity assay, 31
        Infectious titer assay
        Replication-component AAV
            (rcAAV) assay, 32
    genome titer assay, 33, 37, 47
    particle, infectious ratio, 33, 44
    capsid titer assay, 33
    contaminating helper virus assay,
        33, 40
    transgene expression assay, 34
AAV concentration,
    28, 30–31, 72, 139, 143
AAV ITRs, 23
AAV P5 promoter, 22
AAV purification, 24, 26,
    28, 30–31, 33, 42, 44,
AAV Rep proteins, 2, 8, 21–24
AAV serotype vectors
    AAV1, 20, 68, 132,
        168, 193, 196, 204,
        208, 230–231, 264
    AAV2, 1, 20–24, 29–30, 33, 64–66,
        68, 71, 90, 168, 170, 196, 204,
        208, 256, 264, 265

AAV3, 20–21
AAV4, 20–21, 193
AAV5, 20–21, 68, 193, 196, 204
AAV9, 231
AAV vector production, 4, 23,
    44, 64, 163
AAVS1 site, 13, 169
Activated partial thromboplastin
    time, 73
Acute renal failure, 161, 181
Adeno-associated viral vector, 60–61,
    108, 113, 167–168, 178, 182,
    255–256, 258
Adeno-associated virus, 1, 19,
    59, 64, 110, 162, 167–168,
    226, 229, 256
Adenoviral, 3, 22, 25, 60–61,
    113, 143, 167–168, 173,
    175, 178, 182
Adenovirus, 4, 11, 20–21, 25, 32,
    40, 59–60, 108, 113, 143,
    145, 167, 170, 173, 184, 196,
    226–227, 234
Airways, 8, 83–84, 87–88, 90–91, 94
Allograft, 127, 129, 133, 135, 137,
    140–141, 165, 183–184
    rejection, 127–128, 131, 133, 136,
    146–147, 172
Alpha-1 antitrypsin (AAT), 5–7, 62,
    84, 86, 168, 258
Alport syndrome, 161, 176–177
Alveoli, 83, 85
Alzheimer's disease (AD), 198–199,
    204, 206, 208, 212
Amyotrophic lateral sclerosis
    (ALS), 197

Angiogenesis, 103–104, 107, 109, 112–113, 119, 226
Anti-apoptotic gene 20 (A20), 135, 141, 182
Antigen presenting cells (APC), 131
Antigens, 72, 128–129
Anti-inflammatory, 85, 91, 140, 182
AP-1, 134
ApoE, 132
Apoptosis, 109, 130, 134, 140, 182
Aquaporin, 167–168, 178
Aromatic amino acid decarboxylase (AADC), 203
Autoantibodies, 128, 206
Autoimmunity, 126–129, 136, 146
  recurrence, 127–129, 131, 136, 146, 147

B

B Lymphocytes, 131, 138
B7 (B7-1, B7-2), 136
Bcl-2, 135, 141, 182
Bcl-x_L, 135, 141
Beta amyloid protein, 206
β cell, 126, 128–133, 137–138, 140–141, 145–146
  allograft, 129
  destruction of, 130
  reactive T cells, 138
  replacement of, 126
Beta glucuronidase, 12, 199, 209–210, 211
Bethesda assay, 73,
Blood brain barrier, 195, 210, 245, 262
Bone morphogenic protein receptor II, 93

C

Cap gene(s), 13, 23
Capsid, 11–13, 21–22, 24, 31, 33, 40, 43, 47, 69, 84, 91, 170, 174, 179, 208, 212, 230, 256
Carbonic anhydrase, 178
Cardiomyocytes, 228, 229
Cardiovascular disease, 125, 127, 225–226, 234
Carotid artery, 227, 232, 237
Catalase, 131, 135, 141
CD3, 129
CD4, 128–129, 137, 138
CD8, 128–130
CD28, 136
CD80, 136
CD86, 136
CD154, 129
Central nervous system, 195–196, 209
Ceroid lipofuscinosis, 252, 256
CFTR, 3, 4, 7, 88–89
Choline acetyltransferase (Ach), 205
Chronic rejection, 183
Clinical manufacturing, 38, 41, 44–45, 48
Coagulation cascade, 58
Cryoprecipitate, 69
Cryoprecipitation, 72
Current good manufacturing practices (cGMP), 39, 46
Cu-Zn SOD, 135, 141
Cystic Fibrosis (CF), 4, 21, 83–84, 88–89, 168, 266
Cytokine(s), 8, 11, 85, 89, 92, 128, 130, 137, 140, 165, 182
  anti-inflammatory, 85, 140, 182
  T helper 1 (Th1), 92, 130, 137, 138

T helper 2 (Th2), 137
T helper 3 (Th3), 130
Cytoprotective, 132, 140
Cytotoxic T lymphocyte antigen 4 (CTLA-4), 137
  fusion protein (CTLA-4Ig), 137

D

Dendrite cells, 131
Diabetes, type-1, 125, 128–130, 136, 138–139, 142, 144–145, 147
  as health problem, 125,
  β cell destruction in, 128, 129
  cytokines and, 130
  gene therapy for prevention and treatment of, 136, 142, 145, 146
  insulin-dependent (IDDM), 125
  overt, 129
  prevention of autoimmunity recurrence in, 136
  transfer, 129
Diabetic kidney disease, 161, 172, 179, 180
Diabetogenesis, 129

E

Ependymal cells, 195
Epithelium, 9, 84, 88–89, 93, 104, 108–109

F

Fabry, 176–177, 244, 249, 261
  disease, 176–177, 244–245
Factor VIII, 57–65, 67–68, 71, 72
Factor IX, 5, 6, 8, 57, 59, 61–65, 68, 71, 72
Fas, 130, 141
  ligand (FasL), 130, 135, 141
FDA Centre for Biologics Evaluation and Research (CGER), 35,
FKBP52, 90–91
Furin, 143

G

Galanin, 199, 207–208
Gaucher, 209, 244, 249, 256
  disease, 209, 244
Gene Therapy, 1, 2, 36, 40, 57–60, 63, 65, 69–71, 84, 87, 89, 91, 93–94, 113, 129, 131, 136, 139, 142–146, 161, 163, 176, 179, 183–184, 193, 195, 198, 203, 205–207, 210, 225–226, 229, 234, 245, 248, 254–256, 264–265
Glial cell line-derived neurotrophic factor (GDNF), 199, 200, 203, 204
Glomerulus, 162, 176, 181
Gluconeogenic, 143
  pathway, 143
  promoter, 143

Glucose 6-phosphatase (G6P), 143
Glucose, 125, 143, 144, 145
  blood levels, 125, 145
  regulation, 143
  sensing, 125
  transporter–2 (GLUT–2), 143
Glutathion peroxidase, 134, 140, 141
Glycogen storage disease type II, 244, 248
Glycosoaminoglycan, 212
Golgi Apparatus, 143, 253
Granzymes, 130
GTP cyclohydrolases (GTPCH), 200, 202

H

Helper Virus (es), 1, 3, 4, 21, 23, 25–26, 33–34, 40, 170
Heme oxygenase, 135, 182–184
Hemophilia clinical trials, 5, 19, 57, 58, 59, 60, 64
  non-viral vectors, 59, 60, 63
  adenovirus, 60
  retrovirus, 57, 59, 62, 167, 181, 227, 233, 249, 250, 251
  adeno-associated virus, 64–65
Hepatocyte(s), 4, 6, 8–10, 94, 142–144
  growth factor, 164
Human infection, 1–3, 8, 10, 12, 23, 168
Hyperglycemia, 144–145
Hypoglycemic episodes, 126

I

IDDM, 125; *see also* Diabetes
IL–1β, 131, 135, 141

IL–2, 130, 137
IL–4, 92, 130, 133, 138, 139
IL–5, 137
IL–6, 137
IL–10, 92, 130, 133–134, 137–139
IL–12, 134, 138
IL–13, 92, 137
Immunomodulation, 126, 133, 137
Immunoregulation, 136, 137
Inhibitory antibodies, 61, 69, 70, 73
Injury, 84, 94, 134, 140, 163, 166, 169–170, 173, 179–180, 182, 183
  oxidative, 140
In-process sampling, 39, 46
Insulin, 125, 126, 131, 138, 139, 142–146, 179
  autoantibody, 139
  gene therapy, 142, 144–145,
  replacement, 125, 142
  single chain analogue, 143
Insulinoma cells, 134–135
Insulitis, 129–131, 133, 135, 138–139
Integration, 2, 3, 6–10, 12, 13, 21–24, 59, 63, 89, 94, 169, 227, 232, 236
Intercalated cells, 169, 170, 174, 178
Interferon-γ (IFN-γ), 92, 130
Interleukin 10, 182–183
Internal ribosome entry site (IRES), 132
Interstitial cells, 165, 167, 175
Inverted terminal repeat (ITR), 2–3, 23–24, 67, 256
Islet, 12, 126, 129, 132, 136, 140–141, 169, *see also,* Pancreatic islet

K

K-cells, 142–143
Krabbe disease, 209

Kringle domains 1 through 3 of angiostatin (K1K3), 109–113, 119

L

Latency, 2, 9, 23, 169
L-DOPA, 201–203
Leading strand synthesis, 9
Lentivirus, 59, 63, 167, 228–229, 233
Liposomes, 89, 165–166, 173, 178, 232
Low-density lipoprotein receptor (LDL-R), 132
L-pyruvate kinase (L-PK), 143
Lymphocytes, 92, 131, 133, 136–138, 184, 229
Lysosomal storage disorder, 176, 196, 208
Lysosome, 209, 210, 243, 247, 254
  Lysosomal enzymes, 177, 208, 211, 243, 247, 255
  Lysosomal storage diseases, 176, 196, 199, 208, 244, 249, 265

M

Macrophage(s), 128, 131, 137–138, 140, 163
Major histocompatibility complex (MHC), 127
Manganese superoxide dismutase (MnSOD), 134, 140
Mesothelioma, 93
Metachromatic leukodystrophy, 244, 251, 265
MPS I, 244, 250, 256, 265
MPS II, 244, 250, 257

MPS III, 244, 250, 257, 265
MPS VI, 244, 251
MPS VII, 209–211, 251, 263, 265
mRNA, 4, 21, 67, 132, 164, 208

N

Naked DNA, 89, 163–164, 182, 232
Nanoparticles, 165, 232
Natural Killer (NK), 131
Neovascularization, 103–106, 108–109, 113, 117, 234
Nerve growth factor, 205
NeuroD (BETA2), 144, 145
Neuropeptide Y (NPY), 208
NFkB, 134–135
Niemann-Pick disease, 208, 251, 256
NOD Mouse/Mice, 65, 67, 128–131, 133–134, 137–139
  NOD–SCID, 129, 133

O

Oxidative, 131, 134, 140
  damage, 134
  injury, 140
  stress, 131

P

Pancreas, 126, 127, 133, 137–138, 144
Pancreatic duodenal homeobox-1 (Pdx-1 or lpf1), 144
Pancreatic islet, 12, 126, 129, 132, 136, 140–141, 169

Pancreatic islet (*cont.*)
   neogenesis, 126, 142, 144–146
   transduction of, 132, 136
   transplantation, 126–127, 131–132, 136–138, 142, 146
Parkinson disease, 5, 198
Perforin, 130
Pigment epithelium derived factor (PEDF), 104, 108–109
Pleura, 85, 87, 92, 93
Polycystic kidney disease, 161, 176
Pre-clinical compliance, 35, 38, 44, 71, 87, 90
   Good Laboratory Practices (GLP), 36–38
Producer cell lines, 25–26
Productive phase, 2, 39, 48
Proinsulin, 145
Proteasome inhibitors, 91, 93
Pulmonary hypertension, 92–93

Q

Quality assurance unit, 37–38, 48
Quality control, 40, 44, 46, 48–49

R

Receptors, 8, 11–13, 21, 27, 90–91, 106, 141, 162, 169–170, 179, 201, 208, 247
Recombinant adeno-associated viral/virus (rAAV), 19, 89, 145, 162, 168, 173, 177, 229, 256
Red nucleus, 197
Rejection, 127–128, 131, 133, 136, 146–147, 172, *see also* Allograft
Renal fibrosis, 164–165, 180
Renal proximal tubular cells, 167
Rep, 2–3, 8, 12–13, 21–25, 32, 236
Replication, 1–3, 22, 25–26, 32, 43, 62, 110, 170, 198, 227
Retrograde transport, 196, 263, 264
Retrovirus, 57, 59, 62, 143, 167, 181, 226, 228, 233, 249

S

Safety testing, 34, 36–37, 44, 49
   mycoplasma testing, 35, 37, 44, 47
   endotoxin testing, 35, 37, 44, 47
   sterility testing, 35, 37, 44, 47–48
   general safety testing, 35
Second strand synthesis, 91, 170
Sendai virus, 166
Serotypes, 12, 20–21, 24, 84, 91, 93, 132, 168, 170, 179, 193, 195, 204, 212, 230, 256, 258, 264–265
Smad7, 165, 180
Stability program, 34, 36–37, 39, 44
Stem cells, 63, 104, 142, 146, 229, 234

T

T cells, 128–129, 131, 133, 136–137, 140
   activation, 133, 136, 184
      interference, 136
   autoreactive, 128, 140
   $CD4^+$, 128–129, 137, 138
   $CD8^+$, 128–130
   function, 136
   immunoregulatory, 130
   natural killer (NK), 138
   receptor, 130, 136
   regulatory, 128

Tay-Sachs disease, 252, 256
Technology transfer, 39
TGF-β, 130, 134, 139, 180
Th1, 92, 130, 137, 138, *see also* Cytokine(s)
Th2, 137, *see also* Cytokine(s)
Th3, 130, *see also* Cytokine(s)
Thioredoxin, 134, 141
Transcriptional factors, 144, 146
Transgene(s), 3, 5, 7, 9, 12–13, 23–24, 32–34, 59–60, 62, 64, 66, 70–71, 84, 94, 132, 133, 136, 140, 145, 164–165, 167, 170, 179, 182, 196, 206, 226, 260
Transgenic, 130, 133, 138–139, 142–143, 206, 258
Transplant rejection, 133, 146, 161
Transplantation, 126, 127, 132, 136, 142, 146, 163, 176, 181, 183, 244
Tumor necrosis factor-α (TNF-α), 131, 135, 140, 141
Tyrosine hydroxylase (TH), 201–202

U

Ultrasound, 165, 183, 238

V

Vascular endothelial cells, 117, 174, 177
Vascular endothelial growth factor (VEGF), 104, 106, 107, 109, 234
Vasculature, 92, 103, 110–111, 115, 119, 162, 167–168, 173, 177, 203, 233
Vasculogenesis, 103–104, 109
vIL-10, 133, 181
VP1; 2, and 3, 2, 21, 31